Textbook of

Engineering Mathematics II

for Third Semester Diploma in Engineering | Polytechnic Students

Textbook of

Engineering Mathematics II

for Third Semester Diploma in Engineering | Polytechnic Students

A Ganesh MSc MPhil PGDCA
Assistant Professor
Department of Mathematics
The Oxford College of Engineering
Hosur Road, Bommanahalli, Bengaluru 560068
Karnataka

G Balasubramanian MSc MPhil PhD
Associate Professor
Department of Mathematics
Government Arts College (Men)
Krishnagiri, Tamil Nadu

CBS

CBS Publishers & Distributors Pvt Ltd

New Delhi • Bengaluru • Pune • Kochi • Chennai

Textbook of Engineering Mathematics II

ISBN: 978-81-239-1910-2

First Edition: 2010 (2011)

Published by Satish Kumar Jain and produced by Vinod K. Jain for
CBS Publishers & Distributors Pvt Ltd
4819/XI Prahlad Street, 24 Ansari Road, Daryaganj,
New Delhi 110 002, India.
Ph: 23289259, 23266861/67 Fax: +91-011-23243014
Website: www.cbspd.com
e-mail: delhi@cbspd.com
cbspubs@vsnl.com
cbspubs@airtelmail.in

Branches

- Bengaluru: Seema House 2975, 17th Cross, K.R. Road, Banasankari 2nd Stage, Bengaluru 560 070, Karnataka
  Ph: +91-80-26771678/79 Fax: +91-80-26771680 e-mail: bangalore@cbspd.com
- Pune: Bhuruk Prestige, Sr. No. 52/12/2+1+3/2 Narhe, Haveli (Near Katraj-Dehu Road Bypass), Pune 411 051, Maharashtra
  Ph: 020-32404169 e-mail: pune@cbspd.com
- Kochi: 36/14 Kalluvilakam, Lissie Hospital Road, Kochi 682 018, Kerala
  Ph: +91-484-4059061-65 Fax: +91-484-4059065 e-mail: cochin@cbspd.com
- Chennai: 20, West Park Road, Shenoy Nagar, Chennai 600 030, Tamil Nadu
  Ph: +91-44-26260666, 26208620 Fax: +91-44-45530020 email: chennai@cbspd.com

Printed at India Binding House, Noida, UP

to

my dear parents

Shiridi Sai Baba

my dear loving son Monish Sri Sai G

and my wife Mamatha

— A Ganesh

—— *and* ——

my dear parents

and my wife S Geetha

— G Balasubramanian

Foreword

THE OXFORD COLLEGE OF ENGINEERING

(Recognized by Govt. of Karnataka, Affiliated to Visvesvaraya Technological University & Approved by A.I.C.T.E. New Delhi & Accredited by National Board of Accreditation)

Bombmanahalli, Hosur Road, Bangalore – 560 068
91–80–30219601, 30219610 Fax: 91-80-25730551
URL: www.theoxford.edu

A sound knowledge of engineering mathematics is a 'sine qua non' for the modern engineer to attain new heights in all aspects of engineering practice. This book, based on numerous worked examples and interactive problems, is ideal for students from a wide range of academic backgrounds, and can be worked through at the student's own pace. Basic mathematical theories are explained in the simplest of terms, supported by practical engineering examples and applications from a wide variety of engineering disciplines, to ensure the reader can relate the theory to actual engineering practice. This extensive and thorough topic coverage makes this an ideal text for a range of university degree modules. This book caters specifically to the engineering mathematics syllabus of any university in the country. This textbook is an outcome of more than 15 years of teaching experience of the author Mr A Ganesh, Assistant Professor, Department of Mathematics, The Oxford College of Engineering, Bangalore. I am hopeful that this 'new' exhaustive book will be useful to both the students as well as the teachers.

Dr Nagaraj R
Vice Principal
The Oxford College of Engineering
Bangalore - 68

Preface

The *Textbook of Engineering Mathematics II* is designed as a self-contained, comprehensive classroom text for the third semester Diploma in Engineering | Polytechnic students. It is written according to the latest syllabus covering all the universities in India. The topics included are • Integration • Definite Integrals and Applications • Numerical Integration • Differential Equations.

It gives us great pleasure in presenting this book. In this edition, the modifications have been dictated by the changes in the syllabi, which have been appreciated by the teachers, lecturers and students at large using the book in classroom conditions.

The main consideration in writing the book was to present the major requirements of the syllabus in as simple manner as possible. This will help students gain confidence in problem-solving.

Each unit is treated in a systematic and logical presentation of solved examples is followed by an exercise section forming an integral part of the text in which students will get enough questions for practice.

The book is designed as a self-contained, comprehensive and friendly text from students' point of view. Both theory and problems have been explained by using elegant diagrams wherever necessary.

Several solved examples which are important from conceptual as well as examination point of view are given at proper places. At the end of each section, graded problems are added. We hope this will serve the purpose of users in its best capacity.

A Ganesh
G Balasubramanian

Acknowledgements

It gives us great pleasure to present this book *Engineering Mathematics II* for diploma in Engineering/polytechnic students as per the latest syllabus for all Indian universities.

Let us take this opportunity to thank one and all who have actually given us all kinds of support directly and indirectly for bring out the textbook.

We whole heartedly thank our Chairman Mr S Narasaraju Garu, Executive Director Mr SNLV Narasimha Raju Garu, Principal Dr T Krishnan, Vice Principal Prof R Nagaraj, and Head of Department of Mathematics Dr K Mallikarjun, The Oxford College of Engineering, Bangalore.

We thank our well wisher Mr K Venkata Ramana Rao ME, Lecturer, Department of Electrical and Electronics Engineering, MBTS Govt Polytechnic, Guntur, Andhra Pradesh, Prof C Rangaraju, Dr S Murthy and colleagues of the Mathematics Department, Govt Arts College (Men), Krishnagiri, Tamil Nadu.

We would like to thank Mrs, Lalitha Kumari Head Mistress, Anantapur, V Manjunath BM Industries, Bangalore, Mr Balasundar, Mr Nasanakota Jagadeesh Babu, Project Manager Valdel, Bangalore, Mr Rajoli Bhaskar Scientist 'B' National Informatics Center, Bangalore. Dr Gutllapalli Narendra Jayanth (Tinku) BDS Anantapur. Mr Rajoli Chakrapani Assistant Manager Bharath Biotech Ltd, Hyderabad. Mr S Lokesh, Software Engineer, Blue Star, Bangalore, Mr Rajoli Girinath Siebel Consultant IBM India Pvt Ltd., Bangalore, Mr S Dattatreya, Networking Engineer, MPhasis, Bangalore for the assistance they provided at all levels for bringing out this textbook successfully.

We are also grateful to our benevolent Mr Krishnamoorthy, Mr Narayanaswmy (Nani), Vasu and Sreeram Bangalore.

We would also like to thank our family members for the support they provided us through our entire life.

Very special thanks goes out to Mr KR Venkataraj and Bros and well wisher friend N Aswathanarayana Setty, Father-in-law Mr D Srinivas Murthy without whose motivation and encouragement, this would not have possible.

We express our sincere gratitude to CBS Publishers & Distributors, Managing Director Mr SK Jain, Senior Director/Chief Editor Mr YN Arjuna and Bengaluru Division Regional Manager Mr Deepak Rao were the main initiators of this textbook for their suggestions and provisions of the font materials evaluated in this study.

We would also like to thank our friends and students for exchanges of knowledge, skills during our course of time writing book.

A Ganesh
G Balasubramanian

Contents

List of Some Useful Formulae

I. Formulae of Trigonometry and Geometry

1. $\sin(A \pm B) = \sin A \cos B \pm \cos A \sin B$

2. $\cos(A \pm B) = \cos A \cos B \mp \sin A \sin B$

3. $\tan(A \pm B) = \dfrac{\tan A \pm \tan B}{1 \mp \tan A \tan B}$

4. $\sin 2A = 2\sin A.\cos A$

5. $\cos 2A = \cos^2 A - \sin^2 A$

6. $\tan 2A = \dfrac{2\tan A}{1 - \tan^2 A}$

7. $\sin 3A = 3\sin A - 4\sin^3 A$

8. $\cos 3A = 4\cos^3 A - 3\cos A$

9. $\tan 3A = \dfrac{3\tan A - \tan^3 A}{1 - 3\tan^2 A}$

10. $\sin hx = \dfrac{e^x - e^{-x}}{2}$

11. $\cos hx = \dfrac{e^x + e^{-x}}{2}$

12. $\sin x = \dfrac{e^{ix} - e^{-ix}}{2i}$

13. $\cos x = \dfrac{e^{ix} + e^{-ix}}{2}$

14. $\sin C + \sin D = 2\sin\dfrac{C+D}{2}.\cos\dfrac{C-D}{2}$

15. $\sin C - \sin D = 2\cos\dfrac{C+D}{2}.\sin\dfrac{C-D}{2}$

16. $\cos C + \cos D = 2\cos\frac{C+D}{2}\cos\frac{C-D}{2}$

17. $\cos C - \cos D = (-2)\sin\frac{C+D}{2}\sin\frac{C-D}{2}$

18. Area of the ellipse $= \pi ab$

19. Area of the circle $= \pi r^2$

20. Volume of the sphere $= \frac{4}{3}\pi r^3$

21. Volume of the cone $= \frac{1}{3}\pi r^2 h$

22. Circumference of the circle $= 2\pi r$

23. Volume of the right circular cylinder $= \pi r^2 h$

24. Total surface area of the cone $= \pi r\,[r + l]$, $(l =$ slant height$)$

25. Surface area of the sphere $= 4\pi r^2$

II. Formulae of Differential Calculus

1. $y = x^n \quad \therefore \frac{dy}{dx} = nx^{n-1}$

2. $y = \log x \quad \therefore \frac{dy}{dx} = \frac{1}{x}$

3. $y = e^x \quad \therefore \frac{dy}{dx} = e^x$

4. $y = \sin x \quad \therefore \frac{dy}{dx} = \cos x$

5. $y = \cos x \quad \therefore \frac{dy}{dx} = -\sin x$

6. $y = \tan x \quad \therefore \frac{dy}{dx} = \sec^2 x$

7. $y = \text{cossec}\, x \quad \therefore \frac{dy}{dx} = -\text{cosec}\, x.\cot x$

8. $y = \sec x$ $\quad \therefore \frac{dy}{dx} = \sec x . \tan x$

9. $y = \cot x$ $\quad \therefore \frac{dy}{dx} = -\operatorname{cosec}^2 x$

10. $y = \sin^{-1} x$ $\quad \therefore \frac{dy}{dx} = \frac{1}{\sqrt{1-x^2}}$

11. $y = \cos^{-1} x$ $\quad \therefore \frac{dy}{dx} = \frac{-1}{\sqrt{1-x^2}}$

12. $y = \tan^{-1} x$ $\quad \therefore \frac{dy}{dx} = \frac{1}{1+x^2}$

13. $y = \operatorname{cosec}^{-1} x$ $\quad \therefore \frac{dy}{dx} = \frac{-1}{x\sqrt{x^2-1}}$

14. $y = \sec^{-1} x$ $\quad \therefore \frac{dy}{dx} = \frac{1}{x\sqrt{x^2-1}}$

15. $y = \cot^{-1} x$ $\quad \therefore \frac{dy}{dx} = \frac{-1}{1+x^2}$

16. $y = \sinh x$ $\quad \therefore \frac{dy}{dx} = \cos hx$

17. $y = \cosh x$ $\quad \therefore \frac{dy}{dx} = \sin hx$

18. $y = \tanh x$ $\quad \therefore \frac{dy}{dx} = \sec h^2 x$

19. $y = \operatorname{cosech} x$ $\quad \therefore \frac{dy}{dx} = -\operatorname{cosec} hx \cot hx$

20. $y = \operatorname{sech} x$ $\quad \therefore \frac{dy}{dx} = -\sec hx \tan hx$

21. $y = \coth x$ $\quad \therefore \frac{dy}{dx} = -\operatorname{cosec} h^2 x$

22. $y = \sinh^{-1} x$ $\quad \therefore \frac{dy}{dx} = \frac{1}{\sqrt{1+x^2}}$

23. $y = \cosh^{-1} x \quad \therefore \dfrac{dy}{dx} = \dfrac{1}{\sqrt{x^2 - 1}}$

24. $y = \tanh^{-1} x \quad \therefore \dfrac{dy}{dx} = \dfrac{1}{1 - x^2}$

25. $y = \text{cosech}^{-1} x \quad \therefore \dfrac{dy}{dx} = \dfrac{-1}{x\sqrt{1 + x^2}}$

26. $y = \text{sech}^{-1} x \quad \therefore \dfrac{dy}{dx} = \dfrac{-1}{x\sqrt{1 - x^2}}$

27. $y = \coth^{-1} x \quad \therefore \dfrac{dy}{dx} = \dfrac{-1}{x^2 - 1}$

III. Formulae of Integral Calculus

1. $\int x^n dx = \dfrac{x^{n+1}}{x+1}$
2. $\int \dfrac{dx}{x} = \log x$
3. $\int e^x dx = e^x$
4. $\int \sin x\, dx = -\cos x$
5. $\int \cos x\, dx = \sin x$
6. $\int \tan x\, dx = \log \sec x$
7. $\int \sec x\, dx = \log(\sec x + \tan x)$
8. $\int \text{cosec}\, x\, dx = \log(\text{cosec}\, x - \cot x)$
9. $\int \cot x\, dx = \log \sin x$
10. $\int \sec^2 x\, dx = \tan x$
11. $\int \text{cosec}^2 x\, dx = -\cot x$
12. $\int \sec x \tan x\, dx = \sec x$

13. $\int \operatorname{cosec} x \cot x \, dx = -\operatorname{cosec} x$

14. $\int \sin h x \, dx = \cos h x$

15. $\int \cos h x \, dx = \sin h x$

16. $\int \sec h^2 x \cdot dx = \tan hx$

17. $\int \operatorname{cosech}^2 x \cdot dx = -\cot hx$

18. $\int \operatorname{sech} x \cdot \tanh x \, dx = -\sec hx$

19. $\int \operatorname{cosech} x \cdot \cot hx \, dx = -\operatorname{cosech} x$

20. $\int \frac{dx}{\sqrt{a^2 - x^2}} = \sin^{-1} \frac{x}{a}$

21. $\int \frac{dx}{\sqrt{x^2 - a^2}} = \cos h^{-1} \frac{x}{a}$

22. $\int \frac{dx}{\sqrt{a^2 + x^2}} = \sin h^{-1} \frac{x}{a}$

23. $\int \frac{dx}{a^2 + x^2} = \frac{1}{a} \tan^{-1} \frac{x}{a}$

24. $\int \frac{dx}{a^2 - x^2} = \frac{1}{2a} \log\left(\frac{a + x}{a - x}\right)$

25. $\int \frac{dx}{x^2 - a^2} = \frac{1}{2a} \log\left(\frac{x - a}{x + a}\right)$

26. $\int \sqrt{a^2 - x^2} \, dx = \frac{x}{2}\sqrt{a^2 - x^2} + \frac{a^2}{2} \sin^{-1} \frac{x}{a}$

27. $\int \sqrt{a^2 + x^2} \, dx = \frac{x}{2}\sqrt{a^2 + x^2} + \frac{a^2}{2} \sin h^{-1} \frac{x}{a}$

28. $\int \sqrt{x^2 - a^2}\,dx = \frac{x}{2}\sqrt{x^2 - a^2} - \frac{a^2}{2}\cosh^{-1}\frac{x}{a}$

29. $\int_0^a f(x)dx = \int_0^a f(a-x)dx$

30. $\int_{-a}^{a} f(x)dx = 2\int_0^a f(x)dx$, if $f(x)$ is even function,

$= 0$ if $f(x)$ is odd function.

1

Methods of Integration

1.0 Introduction

Integral calculus has its origin in finding the areas bounded by curves. Integration are the inverse process of differentiation. It is also called anti derivative.

1.1 Indefinite Integral and Constant of Integration

If $\frac{d}{dx}[F(x)] = f(x)$ or $F'(x) = f(x)$ is called the integral of $f(x)$ w.r.t'. x' and is denoted by $\int f(x)dx = F(x) + C$, since the derivative of a constant is zero. $\frac{d}{dx}[F(x)] = f(x)$ can also be writen as $\frac{d}{dx}[F(x)+c] = f(x)$. Where c is an arbitrary constant.

Thus, $\int f(x)dx = F(x) + c$

Here c is called the constant of integration. Since c is an arbitrary constant the value of the integral is not unique. Therefore, it is called the indefinite integral. The function $F(x)$ is called the integral and $f(x)$ is called the integrand. The symbol $\int$ stands for integration and it was introduced by a German Mathematician Leibniz in 1975.

For example,

We know that $\frac{d}{dx}(\sin x) = \cos x$ $\quad \therefore \int \cos x = \sin x + c$

$\frac{d}{dx}\left(\frac{x^3}{3}\right) = \frac{1}{3} \cdot 3x^2 = x^2$, $\quad$ Hence $\int x^2 dx = \frac{x^3}{3} + c$

$\int x^2 dx = \frac{x^3}{3} + c$, $\quad$ Therefore $\int \frac{1}{x} dx = \log x + c$

Remark: If $\int \frac{1}{x} dx = \log x + c$ then we have $\int f(x)dx = F(x) + c$

i. $\int \frac{d}{dx}\{f(x)\}dx = \text{F}(x) + c = F(x) + c$

ii. $\frac{d}{dx}\{\int F(x)dx\} = \frac{d}{dx}\{f(x) + c\} = f(x)$

Therefore $\frac{d}{dx}\{\int f(x)dx\} = f(x)$

To find the integral of a function we make use of the known results of differentiation. We list below of the formulae on integration.

Some Standard Forms

1. $\int x^n dx = \frac{x^{n+1}}{n+1} + c,\ n \neq 1$

   Since $\frac{d}{dx}\left(\frac{x^{n+1}}{n+1}\right) = \frac{1}{n+1},\ (n+1)\cdot x^n = x^n$

2. $\int e^x dx = e^x + c$

3. $\int a^x dx = \frac{a^x}{\log a} + c \quad (a > 0,\ a \neq 1)$

4. $\int \sin x dx = -\cos x + c$

5. $\int \cos x dx = \sin x + c$

6. $\int \sec^2 x dx = \tan x + c$

7. $\int \text{cosec}^2 x dx = -\cot x + c$

8. $\int \sec x \tan x dx = \sec x + c$

9. $\int \text{cosec}\, x \cot x dx = -\text{cosec}\, x + c$

10. $\int \frac{1}{\sqrt{1-x^2}} dx = \sin^{-1} x + c$

11. $\int \frac{1}{\sqrt{1+x^2}} dx = \tan^{-1} x + c$

12. $\int \frac{1}{x\sqrt{x^2-1}} dx = \sec^{-1} x + c$

13. $\int \sinh x dx = \cosh x + c$

14. $\int \cosh x dx = \sinh x + c$

15. $\int \operatorname{sech}^2 x dx = \tanh x + c$

16. $\int \operatorname{cosech}^2 x dx = -\coth x + c$

17. $\int \operatorname{sech} x \tanh x dx = -\operatorname{sech} x + c$

18. $\int \operatorname{cosech} x \coth x dx = -\operatorname{cosech} x + c$

    Note: $\int 1 dx = x + c$ and $\int 0 dx = c$

1.2 Rules of Integration

Rule 1 If u is a function of x and k any constant then $\int ku\,dx = k\int u\,dx$

For $\frac{d}{dx}[k\int udx] = k\frac{d}{dx}\int udx = ku$ by above remark.

Rule 2 If u and v are function of x then

$$\int (u+v)dx = \int udx + \int vdx$$

$$\frac{d}{dx}[\int udx + \int vdx] = \frac{d}{dx}[\int udx] + \frac{d}{dx}[\int vdx] = u + v$$

$$\therefore\ (u+v)dx = \int udx + \int vdx$$

The rule can be extended to a finite number of functions. Thus the integral of a sum is equal to sum of the integrals.

WORKED EXAMPLES

Example 1 Evaluate $\int (2x^3 + 3x^2 - 4x + 5)dx$

Solution

Given integral

$$= \int (2x^3 + 3x^2 - 4x + 5)dx$$

$$= 2\int x^3 dx + 3\int x^2 dx - 4\int xdx + 5\int 1dx$$

$$= 2\frac{x^4}{4} + 3\frac{x^3}{3} - 4\frac{x^2}{2} + 5x$$

$$= \frac{x^4}{2} + x^3 - 2x^2 + 5x + c$$

Example 2 Evaluate $\int \frac{ax^2 + bx + c}{\sqrt{x}} dx$

Solution

$$I = \int \frac{ax^2 + bx + c}{\sqrt{x}} dx$$

$$= a\int x^{3/2} dx + b\int x^{1/2} dx + c\int x^{-1/2} dx$$

$$= a\frac{x^{5/2}}{5/2} + b\frac{x^{3/2}}{3/2} + c\frac{x^{1/2}}{1/2}$$

$$= \frac{2a}{5}x^{5/2} + \frac{2b}{3}x^{3/2} + 2cx^{1/2} + c$$

Example 3 Evaluate $\int \left(x^3 + 5^x + \frac{6}{x} \right) dx$

Solution

$$= \int x^3 dx + \int 5^x dx + \int \frac{6}{x} dx$$

$$= \frac{x^6}{6} + \frac{5^x}{\log 5} + 6\log x + c$$

Example 4 Evaluate $\int \frac{\sin^2 x + \cos^2 x}{\sin^2 x \cos^2 x} dx$

Solution

$$= \int \frac{\sin^2 x + \cos^2 x}{\sin^2 x \cdot \cos^2 x} dx$$

$$= \int \left[\frac{1}{\cos^2 x} + \frac{1}{\sin^2 x} \right] dx$$

$$= \int \sec^2 x dx + \int \operatorname{cosec}^2 x dx = \tan x - \cot x + c$$

Example 5 Evaluate $\int \frac{1}{1 - \sin x} dx$

Solution

$$= \int \frac{1}{1 - \sin x} \times \frac{1 + \sin x}{1 + \sin x} dx$$

$$= \int \frac{1 + \sin x}{1 - \sin^2 x} dx$$

$$= \int \frac{1+\sin x}{\cos^2 x} dx$$

$$= \int \left[\frac{1}{\cos^2 x} + \frac{\sin x}{\cos^2 x}\right] dx$$

$$= \int \sec^2 x dx + \int \sec x \tan x dx$$

$$= \tan x + \sec x + c$$

Example 6 Evaluate $\int \sqrt{1+\sin 2x}\ dx$

Solution

$$= \int \sqrt{\sin^2 x + \cos^2 x + 2\sin x \cos x}\ dx$$

$$= \int \sqrt{(\sin x + \cos x)^2}\ dx$$

$$= \int (\sin x + \cos x) dx$$

$$= -\cos x + \sin x + c$$

EXERCISES

Integrate the following functions:

1. $4x^3 - 3x^2 + 2x - 5$ $\left[\textbf{Ans}: x^4 - x^3 + x^2 - 5x + c\right]$
2. $2x^2 - 5x + 3$ $\left[\textbf{Ans}: \frac{2}{3}x^3 - \frac{5}{2}x^2 + 3x + c\right]$
3. $\sqrt[3]{x^2}$ $\left[\textbf{Ans}: \frac{3}{5}x^{5/3} + c\right]$
4. $(1-x)\sqrt{x}$ $\left[\textbf{Ans}: \frac{2}{3}x^{3/2} - \frac{2}{5}x^{5/2} + c\right]$
5. $(3x+4)^2$ $[\textbf{Ans}: 3x^3 + 12x^2 + 16x + c]$
6. $\frac{x^3 + 5x^2 - 4}{x^2}$ $\left[\textbf{Ans}: \frac{1}{2}x^2 + 5x + \frac{4}{x} + c\right]$
7. $1 + \frac{1}{x} + \frac{1}{x^2} + \frac{1}{x^3}$ $\left[\textbf{Ans}: x + \log x - \frac{1}{x} - \frac{1}{2x^2} + c\right]$
8. $e^x + \sin x - \cos x$ $[\textbf{Ans}: ex - \cos x - \sin x + c]$
9. $\frac{ax^2 + bx + c}{x}$ $\left[\textbf{Ans}: \frac{a}{2}x^2 + bx + c\log x + c\right]$

10. $\dfrac{4}{1+x^2}+\dfrac{1}{x}-x$ $\left[\textbf{Ans}: 4\tan^{-1}x \log x-\dfrac{x^2}{2}+c\right]$

11. $\dfrac{3}{\sqrt{1-x^2}}+e^x+4$ [**Ans :** $3\sin^{-1}x + e^x + 4x + c$]

12. $(x+1)^2(x-1)$ $\left[\textbf{Ans}: \frac{x^4}{4}+\frac{x^2}{2}+\frac{x^2}{2}-x+c\right]$

13. $3\text{cosec}^2x + 5e^x$ [**Ans :** $-3\cot x + 5e^x + c$]

14. $3^x+x^3+\dfrac{3}{x}$ $\left[\textbf{Ans}: \dfrac{3^x}{\log 3}+\dfrac{x^4}{4}+3\log x+c\right]$

15. $\dfrac{\sec x}{\sec x-\tan x}$ [**Ans :** $\sec x + \tan x + c$]

16. $\tan^2x$ [**Ans :** $\tan x - x + c$]

17. $\cot^2x$ [**Ans :** $-\cot x - x + c$]

18. $\dfrac{\sin^2 x}{1+\cos x}$ [**Ans :** $x - \sin x + c$]

19. $\dfrac{\cos^2 x}{1-\sin x}$ [**Ans :** $x - \cos x + c$]

20. $\dfrac{1}{1+\sin x}$ [**Ans :** $\tan x - \sec x + c$]

21. $\dfrac{1}{1-\cos x}$ [**Ans :** $-\cot x - \text{cosec}x + c$]

22. $\dfrac{1-\cos^2 x}{1+\cos^2 x}$ [**Ans :** $x - x + c$]

23. $\dfrac{4x^2-2x^2+5x-3}{\sqrt{x}}$ $\left[\textbf{Ans}: \sqrt{x}\left(\frac{2}{7}x^2-\frac{4}{5}x^2+\frac{10}{3}x-6+c\right)\right]$

24. $\dfrac{x^2}{x^2+1}$ [**Ans :** $x - \tan^{-1}x + c$]

25. $\dfrac{x^4+2}{x^2+1}$ $\left[\textbf{Ans}: \frac{1}{3}x^3-x+3\tan^{-1}x+c\right]$

26. $\dfrac{\sin x}{1+\sin x}$ [**Ans :** $\sec x - \tan x + x + c$]

1.3 Integration by the Method of Substitution

Sometimes the given integral can be reduced to one of the above standard forms by changing the independent variables by using the substitution

Rule 3 Let $x = \phi(t)$ where *t* is the new independent variable. Then

$$\int f(x) = dx = \int f(x)\frac{dx}{dt}dt = \int f[\phi(t)]\phi'(t)dt$$

Proof

Let $\quad y = \int f(x)dx \qquad\qquad$ so that $\frac{dy}{dx} = f(x)$

Now $\frac{dy}{dt} = \frac{dy}{dx}\frac{dx}{dt} = f(x)\frac{dx}{dt}$

Integrating *w.r.t.* '*t*', we get

$$\int \frac{dy}{dt}dt = \int f(x)\frac{dx}{dt}dt \qquad\qquad \therefore\ y = \int f(x)\frac{dx}{dt}dt$$

Since $\quad x = \phi(t),\ \frac{dx}{dt} = \phi'(t)$

Hence $y = \int f(x)\phi'(t)dt$

The substitute $x = \phi(t)$ depends upon the nature of the given integral and has to be properly chosen.

Particular cases

If $\int f(x)dx = \phi(x) + c$ then

i. $\quad \int f(ax+b)dx \ = \ \frac{1}{a}\phi(ax+b) + c$ and

ii. $\quad \int f(x^n)x^{n-1}dx \ = \ \frac{1}{n}\phi(x^n) + c$

Proof i. substitute $ax + b = t$ then

$$a\frac{dx}{dt} = 1 \text{ or } dx = \frac{1}{a}dt$$

$$\therefore \int f(ax+b)dx = \int f(t)\cdot\frac{1}{a}dt = \frac{1}{a}\int f(t)dt$$

$$= \frac{1}{a}\phi(t) + c = \frac{1}{a}\phi(ax+b) + c$$

ii. $x^n = t$ then $nx^{n-1}\dfrac{dx}{dt} = 1$ or $x^{n-1}dx = \dfrac{1}{n}dt$

$$\therefore \quad \int f(x^n)x^{n-1}dx = \int f(t)\frac{1}{n}dt$$

$$= \frac{1}{n}\int f(t)dt = \frac{1}{n}\varnothing(t) + c = \frac{1}{n}\varnothing(x^n) + c$$

Generalised Standard Forms

1. $\int x^n dx = \dfrac{x^{n+1}}{n+1} + c,\ n \neq 1$

   $\int (ax+b)^n dx = \dfrac{(ax+b)^{n+1}}{a(n+1)} + c,\ n \neq -1$

2. $\int \dfrac{1}{(ax+b)}dx = \dfrac{1}{a}\log(ax+b) + c$

3. $\int e^{ax+b}dx = \dfrac{1}{a}e^{ax+b} + c$

4. $\int \sin(ax+b)dx = -\dfrac{1}{a}\cos(ax+b) + c$

5. $\int \cos(ax+b)dx = \dfrac{1}{a}\sin(ax+b) + c$

6. $\int \sec^2(ax+b)dx = \dfrac{1}{a}\sin(ax+b) + c$

7. $\int \sec(ax+b)\tan(ax+b)dx = \dfrac{1}{a}\sec(ax+b) + c$

8. $\int \operatorname{cosec}(ax+b)\cot(ax+b)dx = -\dfrac{1}{a}\operatorname{cosec}(ax+b) + c$

For Example

1. $\int (2x+3)^3 dx = \dfrac{(2x+3)^{5+1}}{2(5+1)} = \dfrac{1}{12}(2x+3)^6 + c$

2. $\int \dfrac{1}{3x+1}dx = \dfrac{1}{3}\log(3x+1) + c$

3. $\int \sin 4x dx = -\dfrac{1}{4}\cos 4x + c$

4. $\int e^{3x}dx = \dfrac{1}{3}e^{3x} + c$

5. $\int \cos 5x dx = \dfrac{1}{5}\sin 5x + c$, etc

6. $\int \tan x dx = \int \frac{\sec x \tan x}{\sec x} dx = \log \sec x + c$

Since $\frac{d}{dx}(\sec x) = \sec x \tan x$

7. $\int \cot x dx = \int \frac{\cos x}{\sin x} dx = \log \sin x + c$

Since $\frac{d}{dx}(\sin x) = \cos x$

WORKED EXAMPLES

Example 1 Evaluate $\int \sin^2 x dx$

Solution

$$
\begin{aligned}
I &= \int \frac{1}{2}(1 - \cos^2 x) dx \\
&= \frac{1}{2}\left[\int 1 dx - \int \cos 2x dx\right] \\
&= \frac{1}{2}\left[x - \frac{\sin 2x}{2}\right] + c
\end{aligned}
$$

Example 2 Evaluate $\int \cos^2 x dx$

Solution

$$
\begin{aligned}
I &= \int \frac{1}{2}(1 - \cos^2 x) dx \\
&= \frac{1}{2}\left[\int 1 dx + \int \cos 2x dx\right] = \frac{1}{2}\left[x + \frac{\sin 2x}{2}\right] + c
\end{aligned}
$$

Example 3 Evaluate $\int \sin^3 x \, dx$

Solution

$$
\begin{aligned}
\sin 3x &= 3\sin x - 4\sin^3 x \\
\sin^3 x &= \frac{1}{4}(3\sin x - \sin 3x) \\
\int \sin^3 x dx &= \frac{1}{4}\int (3\sin x - \sin 3x) dx \\
&= \frac{1}{4}\left[\int 3\sin x dx - \int \sin 3x dx\right] \\
&= \frac{1}{4}\left[-3\cos x + \frac{\cos 3x}{3}\right] + c
\end{aligned}
$$

Example 4 Evaluate $\int \cos^3 x\,dx$

Solution

$$\cos 3x = 4\cos^3 x - 3\cos x$$

$$\cos^3 x = \frac{1}{4}(3\cos x + \cos 3x)$$

$$\int \cos^3 x dx = \frac{1}{4}\int (3\cos x + \cos 3x)dx$$

$$= \frac{1}{4}\left[\int (3\cos x dx + \int \cos 3x dx\right]$$

$$= \frac{1}{4}\left[3\sin x + \frac{\sin 3x}{3}\right] + c$$

Example 5 Evaluate $\int \cos 3x \cos 2x\,dx$

Solution

We know that

$$\cos A \cos B = \frac{1}{2}[\cos(A+B) + \cos(A-B)]$$

$$\cos 3x \cos 2x = \frac{1}{2}[\cos(3x+2x) + \cos(3x-2x)]$$

$$\int \cos 3x \cos 2x dx = \frac{1}{2}[\cos 5x + \cos x]$$

$$= \frac{1}{2}\int (\cos 5x + \cos x)dx$$

$$= \frac{1}{2}\left\{\frac{\sin 5x}{5} + \sin x\right\} + c$$

Example 6 Evaluate $\int \sin^3 x \cos x dx$

Solution

Substitute $\sin x = t$, so that $\cos x \frac{dx}{dt} = 1$ or $\cos x dx = dt$

Hence, given integral

$$\int \sin^3 x \cos x\,dx = \int t^3 dt = \frac{t^4}{4} = \frac{\sin^4 x}{4} + c$$

Example 7 Evaluate $\int (x^2 - 2x + 3)^5 (x - 1)dx$

Solution

Substitute $x^2 - 2x + 3 = t$, Differentiate w.r.t. 't'

$$(2x-2)\frac{dx}{dt} = 1 \quad \text{or} \quad 2(x-1)dx = dt \quad \text{or} \quad (x-1)dx = \frac{1}{2}dt$$

Hence, given integral

$$\int (x^2 - 2x + 3)^5 (x-1)dx = \int t^5 \frac{1}{2} dt = \frac{1}{2}\frac{t^6}{6}$$

$$= \frac{(x^2 - 2x + 3)^6}{12}$$

Example 8 Evaluate $\int \frac{(\sin^{-1} x)^2}{\sqrt{1-x^2}} dx$

Solution

Substitute $\sin^{-1} x = t$, Differentiate w.r.t. 't'

$$\frac{1}{\sqrt{1-x^2}}\frac{dx}{dt} = 1 \quad \text{or} \quad \frac{1}{\sqrt{1-x^2}} dx = dt$$

Hence given integral

$$\int \frac{(\sin^{-1} x)^2}{\sqrt{1-x^2}} dx = \int t^2 dt = \frac{t^3}{3}$$

$$= \frac{(\sin^{-1} x)^3}{3} + c$$

Example 9 Evaluate $\int \frac{(\tan^{-1} x)^3}{1+x^2} dx$

Solution

Substitute $\tan^{-1} x = t$. Differentiate w.r.t. 't'

$$\frac{1}{1+x^2}\frac{dx}{dt} = 1 \quad \text{or} \quad \frac{1}{1+x^2} dx = dt$$

Hence given integral

$$\int \frac{(\tan^{-1} x)^3}{1+x^2} dx = \int t^3 dt = \frac{t^4}{4} = \frac{(\tan^{-1} x)^4}{4} + c$$

Example 10 Evaluate $\int \frac{\cos\sqrt{x}}{\sqrt{x}}\,dx$

Solution

Substitute $\sqrt{x} = t$. Differentiate w.r.t. 't'

$$\frac{1}{2\sqrt{x}}\frac{dx}{dt} = 1 \quad \text{or} \quad \frac{1}{\sqrt{x}}dx = 2dt$$

Hence given integral

$$\int \frac{\cos\sqrt{x}}{\sqrt{x}}\,dx = \int \cos t \cdot 2dt = 2\int \cos t\,dt$$

$$= 2\sin t = 2\sin\sqrt{x} + c$$

Example 11 Evaluate $\int \tan^4 x\,dx$

Solution

$$I = \int \tan^2 x \tan^2 x\,dx$$

$$= \int \tan^2 x(\sec^2 x - 1)dx$$

$$= \int [\tan^2 x \sec^2 x - \tan^2 x]dx$$

$$= \int \tan^2 x \sec^2 dx - \int \tan^2 x\,dx$$

In the first put $\tan x = t$, $\therefore\ \sec^2 x\,dx = dt$

$$= \int t^2 dt - \int (\sec^2 x - 1)dx$$

$$= \frac{t^3}{3} - \left(\int \sec^2 x\,dx - \int 1dx\right)$$

$$= \frac{\tan^3 x}{3} - (\tan x - x)$$

$$I = \frac{\tan^3 x}{3} - \tan x + x + c$$

Similarly, $\int \cos^4 x\,dx$, $\int \sec^4 x\,dx$, $\int \mathrm{cosec}^4 x\,dx$ can be evaluated.

Example 12 Evaluate $\int \sin^2 x \cos^2 x\,dx$

Solution

Given Integral

$$= \int (\sin x \cos x)^2 dx$$

$$= \int \left(\frac{\sin 2x}{2}\right)^2 dx, \qquad \text{since } 2\sin x \cos x = \sin 2x$$

$$= \frac{1}{4}\int \sin^2 2x\,dx$$

$$= \frac{1}{4}\int \frac{1}{2}(1-\cos 4x)dx \quad \text{since} \quad \sin^2 A = \frac{1}{2}(1-\cos 2A)$$

$$= \frac{1}{8}\left(x - \frac{\sin 4x}{4}\right) + c$$

Example 13 Evaluate $\int \frac{x}{\sqrt{a^2-x^2}}dx$

Solution

Substitute $a^2 - x^2 = t$, Differentiate w.r.t. 't'

$$-2x\frac{dx}{dt} = 1 \quad \text{or} \quad xdx = -\frac{1}{2}dt$$

Hence given integral

$$\int \frac{x}{\sqrt{a^2-x^2}}dx = \int \frac{1}{\sqrt{t}}\left(-\frac{1}{2}\right)dt$$

$$= \int \frac{1}{\sqrt{t}}\left(-\frac{1}{2}\right)dt$$

$$= -\frac{1}{2}\frac{t^{1/2}}{\frac{1}{2}}$$

$$= -\sqrt{t}$$

$$= -\sqrt{a^2-x^2} + c$$

Example 14 Evaluate $\int \frac{1+\tan x}{1-\tan x}dx$

Solution

Given Integral

$$\int \frac{1+\tan x}{1-\tan x}dx = \int \frac{1+\frac{\sin x}{\cos x}}{1-\frac{\sin x}{\cos x}}dx$$

$$= \int \frac{\cos x + \sin x}{\cos x - \sin x}dx$$

put $t = \cos x - \sin x \quad \therefore \quad dt = (-\sin x - \cos x)dx$

$$dt = -(\sin x + \cos x)dt$$

$$-dt = (\sin x + \cos x)dt$$

$$\therefore \int \frac{1+\tan x}{1-\tan x}dx = -\int \frac{1}{t}dt$$

$$= -\log t + c$$

$$= -\log(\cos x - \sin x) + c$$

Example 15 Evaluate $\int \frac{1}{(a^2+x^2)^{3/2}}dx$

Solution

Substitute $x = a\tan\theta$. Differentiate w.r.t. 'θ'

$$\frac{dx}{d\theta} = a\sec^2\theta \quad \text{or} \quad dx = a\sec^2\theta d\theta$$

Hence given integral

$$\int \frac{1}{(a^2+x^2)^{3/2}}dx = \int \frac{a\sec^2\theta}{(a^2+a^2\tan^2\theta)^{3/2}}d\theta$$

$$= \int \frac{a\sec^2\theta}{[a^2(1+\tan^2\theta)]^{3/2}}d\theta \qquad \because 1+\tan^2\theta = \sec^2\theta$$

$$= \int \frac{a\sec^2\theta}{a^3\sec^3\theta}d\theta$$

$$= \int \frac{1}{a^2\sec\theta}d\theta = \frac{1}{a^2}\int \cos\theta\, d\theta$$

$$= \frac{1}{a^2}\sin\theta$$

but $\frac{x}{a} = \tan\theta \quad \therefore \sin\theta = \frac{x}{\sqrt{a^2+x^2}}$

Given Integral

$$\int \frac{1}{(a^2+x^2)^{3/2}}dx = \frac{1}{a^2}\frac{x}{\sqrt{a^2+x^2}} + c$$

Example 16 Prove that

Solution

$$\int \sec x\,dx = \log(\sec x + \tan x) + c = \log\tan\left(\frac{\pi}{4} + \frac{x}{2}\right) + c$$

Given Integral

$$\int \sec x\,dx = \int \frac{\sec x(\sec x + \tan x)}{\sec x + \tan x}dx$$

$$= \int \frac{\sec^2 x + \sec x \tan x}{\sec x + \tan x}dx$$

put $\sec x + \tan x = t \Rightarrow \left(\sec x \tan x + \sec^2 x\right)dx = dt$

$$\therefore \quad \int \sec x dx = \int \frac{dt}{t} = \log t + c$$

$$= \log(\sec x + \tan x) + c$$

Since numerator is the derivative of the denominator.

$$\sec x + \tan x = \frac{1}{\cos x} + \frac{\sin x}{\cos x} = \frac{1 + \sin x}{\cos x}$$

$$= \frac{1 - \cos\left(\frac{\pi}{2} + x\right)}{\sin\left(\frac{\pi}{2} + x\right)} = \frac{1 - \cos\alpha}{\sin\alpha} \qquad \text{where } \alpha = \frac{\pi}{2} + x$$

$$= \frac{2\sin^2\frac{\alpha}{2}}{2\sin\frac{\alpha}{2}\cos\frac{\alpha}{2}} = \tan\frac{\alpha}{2} = \tan\left(\frac{\pi}{4} + \frac{x}{2}\right)$$

Also $\int \sec x\,dx = \log\tan\left(\frac{\pi}{4} + \frac{x}{2}\right) + c$

Example 17 Prove that

Solution

$$\int \text{cosec}\,x\,dx = \log(\text{cosec} - \cot x) + c = \log\tan\left(\frac{x}{2}\right) + c$$

Given Integral

$$\int \text{cosec}\,x\,dx = \int \frac{\text{cosec}\,x(\text{cosec}\,x - \cot x)}{\text{cosec}\,x - \cot x}dx$$

$$= \int \frac{\text{cosec}^2 x - \text{cosec}\,x \cot x}{\text{cosec}\,x - \cot x}dx = \log(\text{cosec}x - \cot x) + c$$

$$\text{cosec}\,x - \cot x = \tan\frac{x}{2}$$

Given Integral also

$$\int \operatorname{cosec} x\,dx = \log \tan\left(\frac{x}{2}\right) + c$$

Example 18 Evaluate $\int \sin^4 x\,dx$

Solution

Let

$$I = \int \sin^4 x\,dx = \int (\sin^2 x)^2 dx$$

$$= \int \left(\frac{(1-\cos 2x)}{2}\right)^2 dx$$

$$= \frac{1}{4}\int (1 - 2\cos 2x + \cos^2 2x)dx$$

$$\left[\because \cos^2 2x = \frac{1+\cos 4x}{2}\right]$$

$$= \frac{1}{4}\int \left[1 - 2\cos 2x + \frac{1+\cos 4x}{2}\right]dx$$

$$= \frac{1}{8}\int [2 - 4\cos 2x + 1 + \cos 4x]dx$$

$$= \frac{1}{8}\int [3 - 4\cos 2x + \cos 4x]dx$$

$$= \frac{1}{8}\left[3x - 4\frac{\sin 2x}{2} + \frac{\sin 4x}{4}\right]$$

$$= \frac{1}{8}\left[3x - 2\sin 2x + \frac{\sin 4x}{4}\right] + c$$

EXERCISES

Integrate w.r.t. '*x*' the following functions

1. $\sin 3x$ $\quad\left[\textbf{Ans:} -\frac{1}{3}\cos 3x + c\right]$
2. $\cos 5x$ $\quad\left[\textbf{Ans:} \frac{1}{5}\sin 5x + c\right]$
3. $\sec^2 2x$ $\quad\left[\textbf{Ans:} \frac{1}{2}\tan 5x + c\right]$
4. cosex2xcot2x $\quad\left[\textbf{Ans:} -\frac{1}{2}\operatorname{cosec} 2x + c\right]$

5. $\cos(1-2x)$ $\left[\textbf{Ans}: -\frac{1}{2}\sin(1-2x)+c\right]$

6. $(2x+3)^4$ $\left[\textbf{Ans}: \frac{1}{10}(2x+3)^5+c\right]$

7. e^{3x} $\left[\textbf{Ans}: \frac{1}{3}e^{3x}+c\right]$

8. $\frac{1}{e^{zx}}$ $\left[\textbf{Ans}: -\frac{1}{2e^{2x}}+c\right]$

9. $\frac{1}{(1-2x)^3}$ $\left[\textbf{Ans}: \frac{1}{4(1-2x)^2}+c\right]$

10. $\frac{1}{\sqrt{5x+1}}$ $\left[\textbf{Ans}: \frac{2}{5}\sqrt{5x+1}+c\right]$

11. $\frac{1}{(1+x)\sqrt{1+x}}$ $\left[\textbf{Ans}: \frac{-2}{\sqrt{1+x}}+c\right]$

12. $\frac{1}{3x+4}$ $\left[\textbf{Ans}: \frac{1}{3}\log(3x+4)+c\right]$

13. $\sin^2 3x$ $\left[\textbf{Ans}: \frac{1}{2}\left(x-\frac{\sin 6x}{6}\right)+c\right]$

14. $\cos^2 2x$ $\left[\textbf{Ans}: \frac{1}{2}\left(x+\frac{\sin 4x}{4}\right)+c\right]$

15. $\sin^3 x\cos^2 x$ $\left[\textbf{Ans}: -\frac{1}{2}\left(\frac{\cos 5x}{5}+\cos x\right)+c\right]$

16. $\cos^4 x\sin^2 x$ $\left[\textbf{Ans}: \frac{1}{2}\left(\frac{\cos 2x}{2}+\frac{\cos 6x}{6}\right)+c\right]$

17. $\cos^6 x\cos^2 x$ $\left[\textbf{Ans}: \frac{1}{2}\left(\frac{\sin 8x}{8}+\frac{\sin 4x}{4}\right)+c\right]$

18. $\sin 5x\sin 3x$ $\left[\textbf{Ans}: \frac{1}{2}\left(\frac{\sin 2x}{2}+\frac{\sin 8x}{8}\right)+c\right]$

19. $\sin^2 2x\cos^2 2x$ $\left[\textbf{Ans}: \frac{1}{8}\left(x-\frac{\sin 8x}{8}\right)+c\right]$

20. $\sin^3 2x$ $\quad \left[\textbf{Ans:} \frac{1}{4}\left(\frac{\cos 6x}{6} - \frac{3\cos 2x}{2}\right) + c\right]$

21. $\cos^3 2x$ $\quad \left[\textbf{Ans:} \frac{1}{4}\left(\frac{\sin 6x}{6} + \frac{3\sin 2x}{2}\right) + c\right]$

22. $\sin^2 x \cos 3x$ $\quad \left[\textbf{Ans:} \frac{1}{4}\left(\frac{2\sin 3x}{3} + \frac{\sin 5x}{5} - \sin x\right) + c\right]$

23. $\sin\frac{x}{2}\cos\frac{x}{2}$ $\quad \left[\textbf{Ans:} \frac{1}{4}\left(\frac{2\sin 3x}{3} + \frac{\sin 5x}{5} - \sin x\right) + c\right]$

24. $4\cos x\cos 3x\cos 5x$ $\quad \left[\textbf{Ans:} \frac{1}{9}\sin 9x + \frac{1}{7}\sin 7x + \frac{1}{3}\sin 3x + \sin x + c\right]$

25. $\dfrac{\sin 2x}{\cos^2 2x}$ $\quad \left[\textbf{Ans:} \frac{1}{2}\sec 2x + c\right]$

26. $\sin^4 x \cos x$ $\quad \left[\textbf{Ans:} \frac{1}{5}\sin^5 x + c\right]$

27. $\cos^3 x \sin x$ $\quad \left[\textbf{Ans:} -\frac{1}{4}\cos^4 x + c\right]$

28. $\sqrt{\tan x}\sec^2 x$ $\quad \left[\textbf{Ans:} \frac{2}{3}\tan^{3/2} x + c\right]$

29. $(2\sin x + 3)4\cos x$ $\quad \left[\textbf{Ans:} \frac{1}{10}(2\sin x + 3)^5 + c\right]$

30. $(2x^2 + 1)^7 x$ $\quad \left[\textbf{Ans:} \frac{1}{32}(2x^2 + 1)^8 + c\right]$

31. $(x^2 - 2x + 3)^4 (x - 1)$ $\quad \left[\textbf{Ans:} \frac{1}{10}(x^2 - 2x + 3)^5 + c\right]$

32. $\dfrac{x}{\sqrt{x^2 + 1}}$ $\quad \left[\textbf{Ans:} \sqrt{x^2 + 1} + c\right]$

33. $\sec^3 x \tan x$ $\quad \left[\textbf{Ans:} \frac{1}{3}\sec^3 x + c\right]$

34. $\dfrac{\sin x}{\sqrt{1 - 3\cos x}}$ $\quad \left[\textbf{Ans:} \frac{2}{3}\sqrt{1 - 3\cos x} + c\right]$

35. $(2-3\tan x)^3 \sec^2 x$ $\left[\textbf{Ans: } -\dfrac{1}{12}(2-3\tan x)^4+c\right]$

36. xe^{x^2} $\left[\textbf{Ans: } \dfrac{1}{2}e^{x^2}+c\right]$

37. $\dfrac{\sin x}{1+\cos^2 x}$ $\left[\textbf{Ans: } -\tan^{-1}(\cos x)+c\right]$

38. $\dfrac{\cos x}{1+\sin^2 x}$ $\left[\textbf{Ans: } \tan^{-1}(\sin x)+c\right]$

39. $\sqrt{\dfrac{\sin^{-1} x}{1-x^2}}$ $\left[\textbf{Ans: } \dfrac{2}{3}(\sin^{-1} x)^{3/2}+c\right]$

40. $\dfrac{(\tan^{-1} x)^4}{1+x^2}$ $\left[\textbf{Ans: } \dfrac{1}{5}(\tan^{-1} x)^5+c\right]$

41. $\dfrac{1}{\cos^{-1} x\sqrt{1=x^2}}$ $\left[\textbf{Ans: } -\log(\cos^{-1} x)+c\right]$

42. $\dfrac{\log x}{x}$ $\left[\textbf{Ans: } \dfrac{1}{2}(\log x)^2+c\right]$

43. $\dfrac{1}{x\log x}$ $\left[\textbf{Ans: } \log(\log x)+c\right]$

44. $\dfrac{\sin\sqrt{x}}{\sqrt{x}}$ $\left[\textbf{Ans: } -2\cos\sqrt{x}+c\right]$

45. $\dfrac{x-1}{x^2-2x+3}$ $\left[\textbf{Ans: } \dfrac{1}{2}\log(x^2-2x+3)+c\right]$

46. $\dfrac{x}{a^2+x^2}$ $\left[\textbf{Ans: } \dfrac{1}{2}\log(a^2+x^2)+c\right]$

47. $\dfrac{e^x}{e^x+1}$ $\left[\textbf{Ans: } \log(e^x+1)+c\right]$

48. $\dfrac{e^x-1}{e^x+1}$ $\left[\textbf{Ans: } x+2\log(1+e^{-x})+c\right]$

49. $\dfrac{x}{1+x^4}$ $\left[\textbf{Ans: } \dfrac{1}{2}\tan^{-1} x^2+c\right]$

50. $\dfrac{1}{e^x + e^{-x}}$ $\left[\textbf{Ans}: \tan^{-1} e^x + c\right]$

51. $\sec^4 x$ $\left[\textbf{Ans}: \tan x + \dfrac{1}{3}\tan^3 x + c\right]$

52. $\cot^4 x$ $\left[\textbf{Ans}: x + \cot x - \dfrac{1}{3}\cot^3 x + c\right]$

53. $\dfrac{\sin 2x}{1+\sin^2 x}$ $\left[\textbf{Ans}: \log(1+\sin^2 x) + c\right]$

54. $\dfrac{1-e^{2x}}{1+e^{2x}}$ $\left[\textbf{Ans}: x - \log(1+e^{2x}) + c\right]$

55. $\dfrac{\sin x \cos x}{\sqrt{1-\sin^4 x}}$ $\left[\textbf{Ans}: \dfrac{1}{2}\sin^{-1}(\sin^2 x) + c\right]$

56. $\sin^2 x \cos^3 x$ $\left[\textbf{Ans}: \dfrac{1}{3}\sin^3 x - \dfrac{1}{5}\sin^5 x + c\right]$

57. $\dfrac{1}{(a^2 - x^2)^{3/2}}$ $\left[\textbf{Ans}: \dfrac{1}{a^2}\dfrac{x}{\sqrt{a^2 - x^2}} + c\right]$

58. $\dfrac{e^x(1+x)}{\cos^2(xe^x)}$ $\left[\textbf{Ans}: \tan(xe^x) + c\right]$

59. $\dfrac{x-1}{\sqrt{1-x^2}}$ $\left[\textbf{Ans}: -\left(\sqrt{1-x^2} + \sin^{-1} x\right) + c\right]$

60. $\dfrac{x^3}{(1+x^2)^2}$ Hint: Put $x^2+1 = t$

$\left[\textbf{Ans}: \dfrac{1}{2}\left[\log\left(x^2+1\right) + \dfrac{1}{x^2+1}\right] + c\right]$

61. $\dfrac{\sin 2x}{2\cos^2 x + 3\sin^2 x}$ Hint : Put $2\cos^2 x + 3\sin^2 x = t$

$\left[\textbf{Ans}: \log(2\cos^2 x + 3\sin^2 x) + c\right]$

62. $\dfrac{x^3 \tan^{-1} x^3}{1+x^6}$ $\left[\textbf{Ans}: \dfrac{1}{6}\left[\tan^{-1} x^3\right]^2 + c\right]$

63. $\dfrac{\sec x \operatorname{cosec} x}{\log \tan x}$ $\left[\textbf{Ans}: \log(\log \tan x) + c\right]$

64. $\dfrac{\sin 2x}{1 + \sin^4 x}$ $\left[\textbf{Ans}: \tan^{-1}(\sin^2 x) + c\right]$

65. $\dfrac{\sin 2x}{\sin^4 x + \cos^4 x}$ $\left[\textbf{Ans}: \tan^{-1}(\tan^2 x) + c\right]$

66. $\tan^3 x$ $\left[\textbf{Ans}: \dfrac{1}{2}\tan^2 x - \log \sec x + c\right]$

67. $\cos^4 x$ $\left[\textbf{Ans}: \dfrac{1}{8}\left[3x + 2\sin 2x + \dfrac{\sin 4x}{4}\right] + c\right]$

1.4 Integrals of Functions Involving $a^2 \pm x^2$ and their Radicals

Some important integrals

1. Prove that $\int \dfrac{dx}{x^2 + a^2} = \dfrac{1}{a}\tan^{-1}\dfrac{x}{a} + c$

Substitute $x = a\tan\theta$ so that $dx = a\sec^2\theta\, d\theta$

$$\therefore \quad \int \frac{dx}{x^2 + a^2} = \int \frac{a\sec^2\theta d\theta}{a^2\tan^2\theta + a^2}$$

$$= \frac{a}{a^2}\int \frac{\sec^2\theta}{\tan^2\theta + 1} d\theta = \frac{1}{a}\int 1 d\theta \quad (1 + \tan^2\theta = \sec^2\theta)$$

$$= \frac{1}{a}\theta + c = \frac{1}{a}\tan^{-1}\frac{x}{a} + c, \text{ where } \theta = \tan^{-1}\left(\frac{x}{a}\right)$$

Examples

1. $$\int \frac{dx}{x^2 + 16} = \int \frac{dx}{x^2 + 4^2} = \frac{1}{4}\tan^{-1}\frac{x}{4} + c$$

2. $$\int \frac{dx}{4^2 + 9} = \frac{1}{4}\int \frac{dx}{x^2 + \frac{9}{4}} = \int \frac{dx}{x^2 + \left(\frac{3}{2}\right)^2}$$

$$= \frac{1}{4}\frac{1}{(3/2)}\tan^{-1}\frac{x}{(3/2)}$$

$$= \frac{1}{6}\tan^{-1}\frac{2x}{3} + c$$

2. Prove that $\dfrac{dx}{x^2 - a^2} = \dfrac{1}{2a}\log\dfrac{x-a}{x+a} + c$ $\quad (x > a > 0)$

By using partial fractions

$$\frac{1}{x^2 - a^2} = \frac{1}{(x-a)(x+a)} = \frac{1}{2a}\left[\frac{1}{x-a} - \frac{1}{x+a}\right]$$

On integrating both sides w.r.t. 'x'

$$\int \frac{1}{x^2 - a^2}dx = \int \frac{1}{2a}\left[\frac{1}{x-a} - \frac{1}{x+a}\right]dx$$

$$= \int \frac{1}{2a}\left[\int \frac{1}{x-a}dx - \int \frac{1}{x+a}dx\right]$$

$$= \frac{1}{2a}[\log(x-a) - \log(x+a)]$$

$$= \frac{1}{2a}\log\frac{x-a}{x+a} + c$$

Examples

1. $$\int \frac{dx}{x^2 - 4} = \int \frac{dx}{x^2 - 2^2} = \frac{1}{2\cdot 2}\log\frac{x-2}{x+2} = \frac{1}{4}\log\frac{x-2}{x+2} + c$$

2. $$\int \frac{dx}{4x^2 - 25} = \frac{1}{4}\int \frac{dx}{x^2 - \left(\frac{25}{4}\right)} = \frac{1}{4}\int \frac{dx}{x^2 - \left(\frac{5}{2}\right)^2}$$

$$= \frac{1}{4}\frac{1}{2(5/2)}\log\frac{x - \left(\frac{5}{2}\right)}{x + \left(\frac{5}{2}\right)} = \frac{1}{20}\log\frac{2x-5}{2x+5} + c$$

3. Prove that $\displaystyle\int \frac{dx}{a^2 - x^2} = \frac{1}{2a}\log\frac{a+x}{a-x} + c$ $\quad (a > x > 0)$

By using partial fractions,

$$\frac{1}{a^2 - x^2} = \frac{1}{(a+x)(a-x)} = \frac{1}{2a}\left[\frac{1}{a+x} - \frac{1}{a-x}\right]$$

On integrating both sides w.r.t. 'x'

$$\int \frac{1}{a^2 - x^2}dx = \int \frac{1}{2a}\left[\frac{1}{a+x} - \frac{1}{a-x}\right]dx$$

$$= \frac{1}{2a}\left[\int \frac{1}{a+x}dx - \int \frac{1}{a-x}dx\right]$$

$$= \frac{1}{2a}\left[\log(a+x) - \log(a-x)\right]$$

$$= \frac{1}{2a}\log\frac{a+x}{a-x} + c$$

1. $\int \frac{1}{3-x^2} = \int \frac{dx}{\left(\sqrt{3}\right)^2 - x^2} = \frac{1}{2\sqrt{3}}\log\frac{\sqrt{3}+x}{\sqrt{3}-x} + c$

2. $\int \frac{1}{9-4x^2} = \frac{1}{4}\int \frac{dx}{\left(\frac{3}{2}\right)^2 - x^2}$

$$= \frac{1}{4}\frac{1}{2(3/2)}\log\frac{(3/2)+x}{(3/2)-x}$$

$$= \frac{1}{12}\log\frac{3+2x}{3-2x} + c$$

4. Prove that $\int \frac{dx}{\sqrt{a^2 - x^2}} = \sin^{-1}\frac{x}{a} + c$

Substituting $x = a \sin\theta$, so that $dx = a \cos\theta \, d\theta$

$$\int \frac{dx}{\sqrt{a^2 - x^2}} = \int \frac{a\cos\theta d\theta}{\sqrt{a^2 - a^2\sin^2\theta}}$$

$$= \int \frac{a\cos\theta d\theta}{\sqrt{a^2(1-\sin^2\theta)}} d\theta$$

$$= \int \frac{a\cos\theta d\theta}{\sqrt{a^2\cos^2\theta}} d\theta$$

$$= \int 1.d\theta = \theta + c$$

$$= \sin^{-1}\frac{x}{a} + c, \text{ where } \theta = \sin^{-1}\left(\frac{x}{a}\right)$$

Example

1. $\int \frac{dx}{\sqrt{16 - x^2}} = \int \frac{dx}{\sqrt{4^2 - x^2}} = \sin^{-1}\frac{x}{4} + c$

2. $\int \frac{dx}{\sqrt{4-9x^2}} = \frac{1}{3}\int \frac{dx}{\sqrt{(4/9)-x^2}}$

$= \frac{1}{3}\sin^{-1}\frac{x}{2/3}+c$

$= \frac{1}{3}\sin^{-1}\frac{3x}{2}+c$

5. Prove that $\int \frac{dx}{\sqrt{a^2+x^2}} = \sinh^{-1}\frac{x}{a}+c = \log\left[\frac{x+\sqrt{a^2+x^2}}{a}\right]+c$

Substitute $x = a\sinh\theta$ so that $dx = a\cosh\theta\, d\theta$

$a^2+x^2 = a^2+a^2\sinh^2\theta$

$= a^2(1+\sinh^2\theta)$

$= a^2\cosh^2\theta$, since $\cosh^2\theta - \sinh^2\theta$

Given integral

$\int \frac{dx}{\sqrt{a^2+x^2}} = \int \frac{a\cosh\theta}{\sqrt{a^2\cosh^2\theta}}d\theta$

$= a\cosh\theta$

$= \int 1 d\theta = \theta + c$ where $\theta = \sin^{-1}\left(\frac{x}{a}\right)$

$= \sinh^{-1}\frac{x}{a}+c$

Also the given integral $= \log\left[\frac{x+\sqrt{a^2+x^2}}{a}\right]+c$

Examples

1. $\int \frac{dx}{\sqrt{5+x^2}} = \int \frac{dx}{\sqrt{\left(\sqrt{5}\right)+x^2}} = \sinh^{-1}\frac{x}{\sqrt{5}}+c$

2. $\int \frac{dx}{\sqrt{9+4x^2}} = \frac{1}{2}\int \frac{dx}{\sqrt{\left(\frac{9}{4}\right)+x^2}}$

$= \frac{1}{2}\int \frac{dx}{\sqrt{(3/2)^2+x^2}}$

$= \sinh^{-1}\frac{x}{3/2}$

$= \sinh^{-1}\frac{2x}{3}+c$

6. Prove that $\int \frac{dx}{\sqrt{x^2-a^2}} = \cosh^{-1}\frac{x}{a} + c = \log\left[\frac{x+\sqrt{x^2-a^2}}{a}\right] + c$

Substituting $x = a\cosh\theta$, so that $dx = a\sinh\theta\, d\theta$

$$\int \frac{dx}{\sqrt{x^2-a^2}} = \int \frac{a\sinh\theta d\theta}{\sqrt{a^2\cosh^2\theta - a^2}}$$

$$= \int \frac{a\sinh d\theta}{\sqrt{a^2(\cosh^2\theta-1)}} d\theta$$

where, $\sinh^2\theta = \cosh^2\theta - 1$

$$= \int \frac{a\sinh\theta}{\sqrt{a^2\sinh^2\theta)}} d\theta$$

$$= \int \frac{a\sinh\theta}{a\sinh\theta} d\theta = \int 1.d\theta = \theta + c$$

$$= \cosh^{-1}\frac{x}{a} + c \qquad \text{where } \theta = \cosh^{-1}\left(\frac{x}{a}\right)$$

Also the given integral $= \log\left[\frac{x+\sqrt{x^2-a^2}}{a}\right] + c$

Examples

1. $\int \frac{dx}{\sqrt{x^2-4}} = \int \frac{dx}{\sqrt{x^2-2^2}} = \cosh^{-1}\frac{x}{2} + c$

2. $\int \frac{dx}{\sqrt{4x^2-1}} = \frac{1}{2}\int \frac{dx}{\sqrt{x^2-\left(\frac{1}{2}\right)^2}}$

$$= \frac{1}{2}\cosh^{-1}\frac{x}{1/2} = \frac{1}{2}\cosh^{-1}2x + c$$

7. Prove that $\int \sqrt{a^2-x^2}\, dx = \frac{a^2}{2}\left[x + \frac{\sin 2x}{2}\right] + c$

substituting $x = a\sin\theta$, so that $dx = a\cos\theta\, d\theta$

$$\int \sqrt{a^2-x^2}\, dx = \int \sqrt{a^2 - a^2\sin^2\theta}\;\; a\cos\theta d\theta$$

$$= \int \sqrt{a^2(1-\sin^2\theta)} a\cos\theta d\theta$$

$$= \int \sqrt{a^2 \cos^2 \theta)} a \cos\theta d\theta$$

$$= \int a^2 \cos^2 \theta d\theta$$

where $\cos^2\theta = \frac{1}{2}(1+\cos 2\theta)$ and $\theta = x \Rightarrow d\theta = dx$

$$= a^2 \int \frac{1}{2}(1+\cos 2x)dx$$

$$= \frac{a^2}{2}\left[\int 1dx + \int \cos 2xdx\right]$$

$$= \frac{a^2}{2}\left[x + \frac{\sin 2x}{2}\right] + c$$

8. Prove that $\int \sqrt{x^2 - a^2}dx = \frac{a^2}{2}\left(\frac{\sinh 2x}{2} - x\right) + c$

Substituting $x = a \cosh\theta$, so that $dx = a \sinh\theta\, d\theta$

$$\int \sqrt{x^2 - a^2}dx = \int \sqrt{a^2 \cosh^2 \theta - a^2} a \sinh\theta d\theta$$

$$= \int \sqrt{a^2(\cosh^2\theta - 1)} a \sinh\theta d\theta$$

$$= \int \sqrt{a^2 \sinh^2 \theta} a \sinh\theta d\theta$$

$$= \int a \sinh\theta\, a \sinh\theta\, d\theta$$

$$= \int a^2 \sinh^2 \theta d\theta$$

where, $\sinh^2\theta = \frac{1}{2}(\cosh 2\theta - 1)$ and $\theta = x \Rightarrow \theta = dx$

$$= \int a^2 \frac{1}{2}(\cosh 2x - 1)dx$$

$$= \frac{a^2}{2}\int(\cosh 2x - 1)dx$$

$$= \frac{a^2}{2}\left(\frac{\sinh 2x}{2} - x\right) + c$$

9. Prove that $\int \sqrt{a^2 + x^2} = \frac{a^2}{2}\left(\frac{\sinh 2x}{2} + x\right) + c$

Substitute $x = a \sinh\theta$ so that $dx = a \cosh\theta\, d\theta$

$$a^2 + x^2 = a^2 + a^2 \sinh^2 \theta$$

$$= a^2(1 + \sinh^2\theta)$$

$= a^2\cosh^2\theta,$ Since $\cosh^2\theta - \sinh^2\theta$

Given integral

$$\int\sqrt{a^2+x^2} = \int\sqrt{a^2\cosh^2\theta}\,a\cosh\theta d\theta$$

$$= \int a\cosh\theta\, a\cosh\theta d\theta$$

$$= \int a^2\cosh^2\theta\, d\theta\text{, where, }\cosh^2\theta = \frac{1}{2}(\cosh 2\theta+1)$$

$$\text{and }\theta = x \Rightarrow d\theta = dx$$

$$= \int a^2\frac{1}{2}(\cosh 2x+1)dx = \frac{a^2}{2}\int(\cosh 2x+1)dx$$

$$= \frac{a^2}{2}\left(\frac{\sinh 2x}{2}+x\right)+c$$

EXERCISES

Integrate the following w.r.t. 'x'

1. $\dfrac{1}{x^2+16}$ $\left[\textbf{Ans}: \dfrac{1}{4}\tan^{-1}\dfrac{x}{4}+c\right]$
2. $\dfrac{1}{x^2-36}$ $\left[\textbf{Ans}: \dfrac{1}{12}\log\dfrac{x-6}{x+6}+c\right]$
3. $\dfrac{1}{25-x^2}$ $\left[\textbf{Ans}: \dfrac{1}{10}\log\dfrac{5+x}{5-x}+c\right]$
4. $\dfrac{1}{\sqrt{x^2+9}}$ $\left[\textbf{Ans}: \sinh^{-1}\dfrac{x}{3}+c\right]$
5. $\dfrac{1}{\sqrt{x^2-4}}$ $\left[\textbf{Ans}: \cosh^{-1}\dfrac{x}{2}+c\right]$
6. $\dfrac{1}{\sqrt{16-x^2}}$ $\left[\textbf{Ans}: \sin^{-1}\dfrac{x}{4}+c\right]$
7. $\dfrac{1}{4x^2+25}$ $\left[\textbf{Ans}: \dfrac{1}{10}\tan^{-1}\dfrac{2x}{5}+c\right]$
8. $\dfrac{1}{4x^2-9}$ $\left[\textbf{Ans}: \dfrac{1}{12}\log\dfrac{2x-3}{2x+3}+c\right]$
9. $\dfrac{1}{16-25x^2}$ $\left[\textbf{Ans}: \dfrac{1}{40}\log\dfrac{4+5x}{4-5x}+c\right]$

10. $\dfrac{1}{\sqrt{16-9x^2}}$ $\left[\textbf{Ans}: \dfrac{1}{3}\sin^{-1}\dfrac{3x}{4}+c\right]$

11. $\dfrac{1}{\sqrt{9x^2+4}}$ $\left[\textbf{Ans}: \dfrac{1}{3}\sinh^{-1}\dfrac{3x}{2}+c\right]$

12. $\dfrac{1}{\sqrt{4x^2-1}}$ $\left[\textbf{Ans}: \dfrac{1}{2}\cosh^{-1}2x+c\right]$

1.5 Integrals of the Form $\int \frac{dx}{a\cos x + b\sin x + c}$

In such cases use the substitution $\tan\frac{x}{2}=t$

Differentiating w.r.t. 't', we get,

$$\frac{1}{2}\sec^2\frac{x}{2}\frac{dx}{dt}=1$$

Hence $$dx = \frac{2dt}{\sec^2\frac{x}{2}} = \frac{2dt}{1+\tan^2\frac{x}{2}} = \frac{2dt}{1+t^2}$$

Also $$\sin x = \frac{2\tan\frac{x}{2}}{1+\tan^2\frac{x}{2}} = \frac{2t}{1+t^2}$$

and $$\cos x = \frac{1-\tan^2\frac{x}{2}}{1+\tan^2\frac{x}{2}} = \frac{1-t^2}{1+t^2}$$

Substituting the values of $\sin x$, $\cos x$ and dx in the given intergral it reduces to the integral of the form $\int \frac{dt}{At^2+Bt+c}$,

WORKED EXAMPLES

Example 1 Evaluate $\int \frac{dx}{5+3\cos x}$

Solution

Substitute $\tan\frac{x}{2}=t$

$$\text{Hence } dx = \frac{2dt}{1+t^2}, \cos x = \frac{1-t^2}{1+t^2}$$

$$\text{Given integral} = \int \frac{\frac{2dt}{1+t^2}}{5+3\left(\frac{1-t^2}{1+t^2}\right)} = \int \frac{2dt}{5(1+t^2)+3(1-t^2)}$$

$$= \int \frac{2dt}{2t^2+8} = \int \frac{dt}{t^2+4} = \frac{1}{2}\tan^{-1}\frac{t}{2}+c$$

$$= \frac{1}{2}\tan^{-1}\left(\frac{1}{2}\tan\frac{x}{2}\right)+c$$

Example 2 Evaluate $\frac{dx}{3\sin x + 4\cos x + 5}$

Solution

Substitute $\tan\frac{x}{2} = 1$ so that

$$dx = \frac{2dt}{1+t^2}, \sin x = \frac{2t}{1+t^2}, \cos x = \frac{1-t^2}{1+t^2}$$

$$\text{Given integral} = \int \frac{\frac{2dt}{1+t^2}}{3\cdot\frac{2t}{1+t^2}+4\left(\frac{1-t^2}{1+t^2}\right)+5}$$

$$= \int \frac{2dt}{6t+4(1-t^2)+5(1+t^2)}$$

$$= 2\int \frac{dt}{t^2+6t+9}$$

$$= 2\int \frac{dt}{(t-3)^2}$$

$$= 2\frac{(t+3)^{-2+1}}{(-2+1)\cdot 1} = \frac{-2}{t+3}$$

$$= \frac{-2}{\tan\frac{x}{2}+3}+c$$

EXERCISES

Integrate the following w.r.t. '*x*'

1. $\dfrac{1}{5+4\cos x}$ $\left[\textbf{Ans}: \dfrac{2}{3}\tan^{-1}\left(\dfrac{1}{3}\tan\dfrac{x}{2}\right)+c\right]$

2. $\dfrac{1}{13+5\cos x}$ $\left[\textbf{Ans}: \dfrac{1}{6}\tan^{-1}\left(\dfrac{2}{3}\tan\dfrac{x}{2}\right)+c\right]$

3. $\dfrac{1}{2+3\sin x}$ $\left[\textbf{Ans}: \dfrac{1}{\sqrt{5}}\log\dfrac{2\tan\dfrac{x}{2}+3-\sqrt{5}}{2\tan\dfrac{x}{2}+3+\sqrt{5}}+c\right]$

4. $\dfrac{1}{4+5\sin x}$ $\left[\textbf{Ans}: \dfrac{1}{3}\log\dfrac{2\tan\dfrac{x}{2}+1}{2\left(\tan\dfrac{x}{2}+2\right)}+c\right]$

5. $\dfrac{1}{1+\sin x+\cos x}$ $\left[\textbf{Ans}: \log\left(1+\tan\dfrac{x}{2}\right)+c\right]$

6. $\dfrac{1}{3+2\sin x-\cos x}$ $\left[\textbf{Ans}: \tan^{-1}\left(1+2\tan\dfrac{x}{2}\right)+c\right]$

7. $\dfrac{1}{\sin x-2\cos x-2}$ $\left[\textbf{Ans}: \log\left(\tan\dfrac{x}{2}-2\right)+c\right]$

8. $\dfrac{1}{5\sin x-4\cos x+3}$ $\left[\textbf{Ans}: \dfrac{1}{4\sqrt{2}}\log\dfrac{7\tan\dfrac{x}{2}+5-4\sqrt{2}}{7\tan\dfrac{x}{2}+5+4\sqrt{2}}+c\right]$

9. $\dfrac{1}{\sin x+7\cos x+5}$ $\left[\textbf{Ans}: \dfrac{1}{5}\log\dfrac{2+\tan\dfrac{x}{2}}{3-\tan\dfrac{x}{2}}+c\right]$

1.6 Integration by Parts

If *u* and *v* are functions of *x* we have

$$\frac{d}{dx}(uv) = u\frac{dv}{dx}+v\frac{du}{dx}$$

Hence by the definition of integration

$$uv = \int\left[u\frac{dv}{dx}+v\frac{du}{dx}\right] = \int u\frac{dv}{dx}dx+\int v\frac{du}{dx}dx$$

$$\therefore \qquad \int\frac{dv}{dx}dx = uv-\int v\frac{du}{dx}dx \qquad \ldots 1$$

Now substitute $u = f(x)\frac{dv}{dx} = g(x)$ so that $\int g(x)$ and $\frac{du}{dx} = f'(x)$

In (1).

Hence

$$\int f(x)g(x)dx = f(x)\int g(x)dx-\int\left[\int g(x)dx\right]f'(x)dx \qquad \ldots 2$$

When the integrand is the product of two functions, we choose one function as $f(x)$ and the other as $g(x)$, Our choice will be proper only if

(1) $g(x)$ is easily integrable, and (2) the second integral on the R.H.S of (2) must not be more complicated then the given integral.

(2) Usually the functions of the form x, x^2, x^3.... are chosen as $f(x)$

Example 1 Evaluate $\int xe^x dx$

Solution

Here choose $f(x) = x$ and $g(x) = e^x$

On integrating by parts, we get

$$\int xe^x dx = x\int e^x dx-\int\left[\int e^x dx\right]\frac{d}{dx}(x)dx = xe^x-\int e^x.1.dx$$

$$= xe^x-e^x = (x-1)e^x+c$$

Example 2 Evaluate $\int x\sin x\,dx$

Solution

Choosing x as the first function and integrating by parts, we get

$$\int x\sin x\,dx = x\int\sin xdx-\int\left[\int\sin xdx\right]\frac{d}{dx}(x)dx$$

$$= x(-\cos x)-\cos x.1dx$$

$$= -x\cos x+\int\cos x\,dx$$

$$= -x\cos x+\sin x+c$$

Example 3 Evaluate $\int x\cos dx$

Solution

$$\int x\cos xdx = x\int\cos xdx-\int\left[\int\cos xdx\right]\frac{d}{dx}(x)dx$$

$$= x\sin x-\int\sin x1dx$$

$$= x\sin x+\cos x+c$$

Example 4 Evaluate $\int x^2 \sin x dx$

Solution

On integrating by parts, we get,

$$
\begin{aligned}
I &= x^2 \int \sin x dx - \int \left[\int \sin x dx \right] \frac{d}{dx}(x^2) dx \\
&= x^2(-\cos x) - \int (\int -\cos x) 2x dx \\
&= -x^2 \cos x + 2\int x \cos x \, dx \\
&= -x^2 \cos x + 2[x \sin x - \int \sin x \, dx]
\end{aligned}
$$

On integrating by parts again.

$$\therefore \quad I = -x^2 \cos x + 2[x \sin x + \cos x] + c$$

Example 5 Evaluate $\int x \cos^2 x \, dx$

Solution

Given Integral

$$= \int x \frac{(1+\cos 2x)}{2} dx \qquad \text{where } \cos^2 x = \frac{1+\cos 2x}{2}$$

$$
\begin{aligned}
&= \frac{1}{2}\int (x + x\cos 2x) dx \\
&= \frac{1}{2}\left[\int x dx + \int x \cos 2x \, dx\right] \\
&= \frac{1}{2}\left[\frac{x^2}{2} + x\int \cos 2x \, dx - \int \{\int \cos 2x \, dx\} 1 dx\right] \\
&= \frac{1}{2}\left[\frac{x^2}{2} + \frac{x \sin 2x}{2} - \int \frac{\sin 2x}{2} dx\right] \\
&= \frac{1}{2}\left[\frac{x^2}{2} + \frac{x \sin 2x}{2} + \frac{\cos 2x}{4}\right] + c
\end{aligned}
$$

Example 6 Evaluate $\int \sqrt{x^2 + a^2} dx$

Solution

Let $\quad I = \int \sqrt{x^2 + a^2} dx$

$$= \int \sqrt{x^2 + a^2} .1. dx$$

On integrating by parts we get,

$$I = \sqrt{x^2+a^2}.x - \int x\cdot\frac{1}{2\sqrt{x^2+a^2}}2x\,dx$$

$$= x\sqrt{x^2+a^2} - \int\frac{x^2}{\sqrt{x^2+a^2}}dx$$

$$= x\sqrt{x^2+a^2} - \int\left\{\sqrt{x^2+a^2} - \frac{a^2}{\sqrt{x^2+a^2}}\right\}dx$$

$$= x\sqrt{x^2+a^2} - \int\sqrt{x^2+a^2}\,dx + a^2\int\frac{1\cdot dx}{\sqrt{x^2+a^2}}$$

$$I = x\sqrt{x^2+a^2} - I + a^2\sinh^{-1}\left(\frac{x}{a}\right)$$

$$2I = x\sqrt{x^2+a^2} + a^2\sinh^{-1}\left(\frac{x}{a}\right)$$

$$I = \frac{x}{2}\sqrt{x^2+a^2} + \frac{a^2}{2}\sinh^{-1}\left(\frac{x}{a}\right) + c$$

Example 7 Evaluate $\int \log x\,dx$

Solution

Given integral $= \int \log x.1.dx$

$$= \log x\int 1dx - \int\left[\int 1dx\right]\frac{d}{dx}(\log x)dx$$

$$= \log x.x - \int x.\frac{1}{x}dx \quad = x\log x - x + c$$

Example 8 Evaluate $\int \tan^{-1} x\,dx$

Solution

Given integral

$$= \int \tan^{-1} x.1dx$$

$$= \tan^{-1} x\int 1dx - \int\left[\int 1dx\right]\frac{d}{dx}(\tan^{-1} x)dx$$

$$= \tan^{-1} x.x - \int x\cdot\frac{1}{1+x^2}dx$$

$$= x\tan^{-1}x - \frac{1}{2}\cdot\int\frac{2x}{1+x^2}dx$$

$$= x\tan^{-1}x - \frac{1}{2}\log(1+x^2) + c$$

Example 9 Evaluate $\int\frac{x\sin^{-1}x}{\sqrt{1-x^2}}dx$

Solution

Substitute $\sin^{-1}x = t$. Differentiate w.r.t. 't'

$$\frac{1}{\sqrt{1-x^2}}\frac{dx}{dt} = 1 \quad \text{or} \quad \frac{1}{\sqrt{1-x^2}}dx = dt, \qquad \text{also } x = \sin t$$

Hence given integral

$$\int\frac{x\sin^{-1}x}{\sqrt{1-x^2}}dx = \int \sin t.t\,dt$$

$$\Rightarrow \quad = \int t.\sin dt = t\int \sin t\,dt - \int\left(\int \sin t dt\right)\frac{d}{dt}(t).dt$$

$$= t(-\cos t) - \int(-\cos t)dt$$

$$= -t\cos t + \sin t$$

$$= -t\sqrt{1-\sin^2 t} + \sin t$$

$$= -\sin^{-1}x.\sqrt{1-x^2} + x + c$$

Example 10 Evaluate $\int e^{ax}\sin bx\,dx$

Solution

Let $\qquad I = \int e^{ax}\sin bx\,dx$

On integrating by parts,

$$I = e^{ax}\left(\frac{-\cos bx}{b}\right) - \int\left(\frac{-\cos bx}{b}\right)ae^{ax}dx$$

$$= \frac{-e^{ax}\cos bx}{b} + \frac{a}{b}\int e^{ax}\cos bx\,dx$$

$$= \frac{-e^{ax}\cos bx}{b} + \frac{a}{b}\left[e^{ax}\frac{\sin bx}{b} - \int\frac{\sin bx}{b}e^{ax}a\,dx\right]$$

On integrating by parts again

$$= \frac{-e^{ax}\cos bx}{b} + \frac{a}{b}\left[e^{ax}\frac{\sin bx}{b} - \frac{a}{b}\int e^{ax}\sin bx\,dx\right]$$

$$= \frac{-e^{ax}\cos bx}{b} + \frac{ae^{ax}\sin bx}{b^2} - \frac{a^2}{b^2}\,\mathrm{I}$$

$$I\left(1+\frac{a^2}{b^2}\right) = e^{ax}\left[\frac{-b\cos bx + a\sin bx}{b^2}\right]$$

$$\therefore \qquad I = \frac{e^{ax}}{a^2+b^2}[-b\cos bx + a\sin bx] + c$$

Example 11 Evaluate $\int \frac{x+\sin x}{1+\cos x}dx$

Solution

Given Integral

$$\int \frac{x+\sin x}{1+\cos x}dx = \int \frac{x}{1+\cos x}dx + \int \frac{\sin x}{1+\cos x}dx$$

$$= \int \frac{x}{2\cos^2\frac{x}{2}}dx + \int \frac{2\sin\frac{x}{2}\cos\frac{x}{2}}{2\cos^2\frac{x}{2}}dx$$

where, $1+\cos x = 2\cos^2\frac{x}{2}$ and

$$\sin x = 2\sin\frac{x}{2}\sin\frac{x}{2}$$

$$= \frac{1}{2}\int x\sec^2\frac{x}{2}dx + \int \tan\frac{x}{2}dx$$

$$= \frac{1}{2}\left[x\int \sec^2\frac{x}{2}dx - \left\{\int \sec^2\frac{x}{2}dx\right\}\cdot 1dx\right] + \int \tan\frac{x}{2}dx$$

$$= \frac{1}{2}\left[x\cdot 2\tan\frac{x}{2} - \int 2\tan\frac{x}{2}dx\right] + \int \tan\frac{x}{2}dx$$

$$= x\tan\frac{x}{2} - \int \tan\frac{x}{2}dx + \int \tan\frac{x}{2}dx$$

$$= x\tan\frac{x}{2} + c$$

1.7 Bernoulli's Rule

Here we give an alternate and simple method for integration by parts. In what we denote the differentiation by dashes and integration by suffixes. In (2), we have proved that,

$$\int f(x)g(x)dx = f(x)g_1(x) - \int g_1(x)f'(x)dx$$

Where $g_1(x) = \int g(x)dx$

On integrating the integral on the R.H.S by parts again, we get,

$$\begin{aligned}\int f(x)g(x)dx &= f(x)g_1(x) - [f'(x)g_2(x) - \int g_2(x)f''(x)dx] \\ &= f(x)g_1(x) - f'(x)g_2(x) + \int f''(x)g_2(x)dx\end{aligned}$$

Continuing in this way we get,

$$\int f(x)g(x)dx = f(x)g_1(x) - f'(x)g_2(x) + \int f''(x)g_3(x) \qquad \ldots 3$$

Where $g_2(x) = \int g_1(x)dx$, $g_3(x) = \int g_2(x)dx$ etc. We notice that the signs are alternately positive and negative and we go on differentiating $f(x)$ and integrating $g(x)$. This result gives immediate evaluation of integrals involving one of the functions as-integral powers of x

Let $f(x) = u, g(x) = v$

$$\therefore \quad \int uvdx = uv_1 - u'v_2 + u''v_3 - u'''v_4$$

This rule is useful when one of the function is algebraic.

For Example

1. $\int xe^x dx = xe^x - 1.x^x + 0$
2. $\int x\sin x dx = x(-\cos x) + \sin x = -x\cos x + \sin x + c$
3. $\int x\sec^2 x dx = x\tan x - 1 = x\tan x - \log\sec x + c$
4. $\int x^2 \cos x dx = x^2 \sin x - 2x(-\cos x) + 2(-\sin x) - 0$
$= x^2 \sin x + 2x\cos x - 2\sin x + c$
5. $\int x^3 e^{2x} dx = x^3 \frac{e^{2x}}{2} - 3x^2 \frac{e^{2x}}{4} + 6x\frac{e^{2x}}{8} - 6\frac{e^{2x}}{16} + 0$
$= \left(\frac{x^3}{2} - \frac{3x^2}{4} + \frac{3x}{4} - \frac{3}{8}\right)e^{2x} + c$
6. $\int x\sin^2 x dx = \int x.\left(\frac{1-\cos 2x}{2}\right)dx$ where, $\sin^2 = \frac{1-\cos 2x}{2}$
$= \frac{1}{2}\left[\int xdx - \int x\cos 2xdx\right]$

$$= \frac{1}{2}\left[\frac{x^2}{2} - \left\{x.\frac{\sin 2x}{2} - 1.\left(-\frac{\cos 2x}{4}\right) + 0\right\}\right]$$

$$= \frac{1}{2}\left[\frac{x^2}{2} - \frac{x\sin 2x}{2} - \frac{\cos 2x}{4}\right] + c$$

WORKED EXAMPLES

Example 1 Evaluate $\int x^2 e^x \, dx$

Solution
By applying Bernolli's formula
we have,

$$\int uv dx = uv_1 - u'v_2 + u''v_3 - u''v_4 + ...$$

Let $u = x^2, \quad v = e^x$

$$u' = 2x, \quad v_1 = \int e^x dx = e^x$$

$$u'' = 2 \quad v_2 = \int v_1 dx = \int e^x dx = e^x$$

$$v_3 = \int v_2 dx = \int e^x dx = e^x$$

$$\therefore \quad \int x^2 e^x dx = x^2 \cdot e^x - 2x \cdot e^x + 2e^x + C$$

$$= (x^2 - 2x + 2)e^x + C$$

Example 2 Evaluate $\int x^3 \sin x dx$

Solution By applying Bernulli's Formula

$$\int uv dx = uv_1 - u'v_2 + u''v_3 - u''v_4 + ...$$

Let $u = x^3, \quad v = \sin x$

$$u' = 3x^2, \quad v_1 = -\cos x$$

$$u'' = 6x \quad v_2 = -\sin x$$

$$u''' = 6 \quad v_3 = +\cos x$$

$$\therefore \quad \int x^3 \sin dx = x^3(-\cos x) - 3x^2(-\sin x) + 6x(\cos x) + 6(\sin x) + C$$

$$= -x^3 \cos x + 3x^2 \sin x + 6x\cos x + 6\sin x + C$$

$$= (6x - x^3)\cos x + (3x^2 + 6)\sin x + C$$

$$= x(6 - x^2)\cos x + 3(x^2 + 2)\sin x + C$$

Example 3 Evaluate $\int x^3 \cos x dx$

Solution By applying Bernuli Formula,

$$\int uvdx = uv_1 - u''v_2 + u'''v_3 - ....$$

Let $u = x^3, \quad v = \cos x$

$$u' = 3x^2, \quad v_1 = \sin x$$
$$u'' = 6x \quad v_2 = -\cos x$$
$$u''' = 6 \quad v_3 = -\sin x$$
$$v_4 = \cos x$$

$$\begin{aligned}\therefore \quad \int x^3 \cos x dx &= x^3 \sin x - 3x^2(-\cos x) + 6x(-\sin x) + 6\cos x + C \\ &= x^3 \sin x + 3x^2 \cos x - 6x \sin x + 6\cos x + C \\ &= x\left(x^2 - 6\right)\sin x + 2(x^2 + 6)\cos x + C\end{aligned}$$

Example 4 Evaluate $\int x^3 \cos^2 x dx$

Solution

Let $u = x^3, \quad v = \cos^2 x$

$$v = \frac{1}{2}(1 + \cos 2x)$$

$$u_1 = 3x^2, \quad v_1 = \frac{1}{2}\left(x + \frac{\sin 2x}{2}\right)$$

$$u'' = 6x \quad v_2 = \frac{1}{2}\left(\frac{x^2}{2} - \frac{\cos 2x}{4}\right)$$

$$u''' = 6 \quad v_3 = \frac{1}{2}\left(\frac{x^3}{6} - \frac{\sin 2x}{8}\right)$$

$$v_4 = \frac{1}{2}\left(\frac{x^4}{24} + \frac{\cos 2x}{16}\right)$$

By applying Bernoulli's Formula:

$$\int uvdx = uv_1 - u'v_2 + u''v_3 - u''v_4 + ...$$

$$\therefore \int x^3 \cos^2 x dx = x^3 \frac{1}{2}\left(x + \frac{\sin 2x}{2}\right) - 3x^2 \cdot \frac{1}{2}\left(\frac{x^2}{2} - \frac{\cos 2x}{4}\right) +$$

$$6x \cdot \frac{1}{2}\left(\frac{x^3}{6} - \frac{\sin 2x}{8}\right) - 6 \cdot \frac{1}{2}\left(\frac{x^4}{24} + \frac{\cos 24}{16}\right)$$

$$= \frac{x^3}{2}\left(x+\frac{\sin 2x}{2}\right)-\frac{3x^2}{2}\left(\frac{x^2}{2}-\frac{\cos 2x}{4}\right)+$$

$$3x\left(\frac{x^3}{6}-\frac{\sin 2x}{8}\right)-3\left(\frac{x^4}{24}+\frac{\cos 2x}{16}\right)+C$$

1.8 Integrals of the Form $\int e^x\{f(x)+f'(x)\}dx$

Consider

$$\int e^x\{f(x)+f'(x)\}dx = \int e^x\{f(x)+e^x f'(x)\}dx$$

$$= \int\{e^x f(x)+e^x f'(x)\}dx$$

$$= \int e^x f(x)dx+\int e^x f'(x)dx$$

$$= \int e^x f(x)dx+e^x\int f'(x)dx-\int\left[\int f'(x)dx\right]\frac{d}{dx}(e^x)dx$$

$$= \int e^x f(x)dx+e^x f(x)-\int f(x)e^x dx$$

For example,

i. $\int e^x(\sin x+\cos x)dx = e^x \sin x + c$.

ii. $\int e^x\left(\log x+\frac{1}{x}\right)dx = e^x \log x + c$

Example 11 Evaluate $\int \frac{xe^x}{(x+1)^2}dx$

Solution

Given Integral

$$\int \frac{xe^x}{(x+1)^2}dx = \int \frac{(x+1-1)e^x}{(x+1)^2}dx = \int\left[\frac{1}{(x+1)}-\frac{1}{(x+1)^2}\right]e^x dx$$

$$= \frac{1}{(x+1)}e^x \quad \text{since } \frac{d}{dx}\left\{\frac{1}{x+1}\right\}=\frac{1}{(x+1)^2}$$

$$= \frac{1}{(x+1)}e^x + c$$

EXERCISES

Integrate w.r.t. 'x'

1. x^2e^x $\left[\textbf{Ans}: (x^2-2x+2)e^x+c\right]$
2. xe^{2x} $\left[\textbf{Ans}: (2x-1)\frac{e^{2x}}{4}+c\right]$

3. $x \sin 3x$ $\left[\textbf{Ans:} \frac{-x\cos 3x}{3} + \frac{\sin 3x}{9} + c\right]$

4. $x \cos 4x$ $\left[\textbf{Ans:} \frac{x\sin 2x}{2} + \frac{\cos 3x}{4} + c\right]$

5. $x^2 \cos x$ $\left[\textbf{Ans:} x^2 \sin x - 2\sin x + 2x\cos x + c\right]$

6. $(2x + 3) \sin 2x$ $\left[\textbf{Ans:} \frac{1}{2}[\sin 2x - (2x+3)\cos 2x] + c\right]$

7. $x \cos (2x - 1)$ $\left[\textbf{Ans:} \frac{1}{2} x\sin(2x-1) + \frac{1}{4}\cos(2x-1) + c\right]$

8. $x \sin^2 x$ $\left[\textbf{Ans:} \frac{1}{2}\left[\frac{x^2}{2} - \frac{x\sin 2x}{2} - \frac{\cos 2x}{4}\right] + c\right]$

9. $x \sin^2 3x$ $\left[\textbf{Ans:} \frac{1}{2}\left[\frac{x^2}{2} - \frac{x\sin 6x}{6} - \frac{\cos 6x}{36}\right] + c\right]$

10 $\frac{x}{e^{2x}}$ $\left[\textbf{Ans:} -(2x-1)\frac{e^{-2x}}{4} + c\right]$

11. $x\sin 3x \cos x$

$$\left[\textbf{Ans:} \frac{1}{4}\left[\frac{1}{8}\sin 4x + \frac{1}{2}\sin 2x - \frac{1}{2}x\cos 4x - x\cos 2x\right] + c\right]$$

12. $\sin^{-1} x$ $\left[\textbf{Ans:} x\sin^{-1} x + \sqrt{1-x^2} + c\right]$

13. $(\log x)^2$ $\left[\textbf{Ans:} x(\log x)^2 - 2x\log x + 2x + c\right]$

14. $x\log x$ $\left[\textbf{Ans:} \frac{x^2}{4}(2\log x - 1) + c\right]$

15. $\frac{\log x}{x^2}$ $\left[\textbf{Ans:} -\frac{1}{x}(1 + \log x) + c\right]$

16. $x \tan^{-1} x$ $\left[\textbf{Ans:} \frac{1}{2}(x^2+1)\tan^{-1} x - \frac{x}{2} + c\right]$

17. $x \sin^{-1} x$ $\left[\textbf{Ans:} \frac{1}{4}\left[x\sqrt{1-x^2} + (2x^2-1)\sin^{-1}\right] + c\right]$

18. $x \cos 3x \cos x$

$$\left[\textbf{Ans:} \frac{1}{32}[4x\sin 4x + \cos 4x + 16x\sin 2x + 4\cos 2x] + c\right]$$

19. $e^{\sqrt{x}}$ $\left[\textbf{Ans:} 2\left(\sqrt{x-1}\right)e^{\sqrt{x}}+c\right]$

20. $\cos\sqrt{x}$ $\left[\textbf{Ans:} 2\left(\sqrt{x}\sin\sqrt{x}+\cos\sqrt{x}\right)+c\right]$

21. $x^n \log x$ $\left[\textbf{Ans:} \frac{x^{n-1}}{n+1}\left[\log x-\frac{1}{n+1}\right]+c\right]$

22. $x \log(1+x)$ $\left[\textbf{Ans:} \frac{x^2}{2}\log(1+x)-\frac{1}{2}\left[\frac{x^2}{2}-x+\log(1+x)\right]+c\right]$

23. $x\sec^2 2x$ $\left[\textbf{Ans:} \frac{1}{2}x\tan 2x-\frac{1}{4}\log\sec 2x+c\right]$

24. $\cot^{-1}x$ $\left[\textbf{Ans:} x\cot^{-1}x+\frac{1}{2}\log(1+x^2)+c\right]$

25. $x\sec^{-1}x$ $\left[\textbf{Ans:} \frac{x^2}{2}\sec^{-1}x-\frac{1}{2}\sqrt{x^2-1}+c\right]$

26. $\sin^{-1}\sqrt{x}$ $\left[\textbf{Ans:} \left(x-\frac{1}{2}\right)\sin^{-1}\sqrt{x}+\frac{1}{2}\sqrt{x-x^2}+c\right]$

27. $x^2\tan^{-1}x$ $\left[\textbf{Ans:} \frac{x^3}{3}\tan^{-1}x-\frac{x^2}{2}+\frac{1}{6}\log(1+x^2)+c\right]$

28. $x^2 a^x$ $\left[\textbf{Ans:} \frac{a^x}{\log a}\left[x^2-\frac{2x}{\log a}+\frac{2}{(\log a)^2}\right]+c\right]$

29. $\dfrac{(x-1)e^x}{x^2}$ $\left[\textbf{Ans:} \frac{e^x}{x}+c\right]$

30. $e^x\cos 2x$ $\left[\textbf{Ans:} \frac{e^x}{5}\left[x^2-\frac{2x}{\log a}+\frac{2}{(\log a)^2}\right]+c\right]$

31. $e^{2x}\sin 2x$ $\left[\textbf{Ans:} \frac{e^{2x}}{13}(2\sin 3x-2\cos 3x)+c\right]$

32. $\dfrac{1+x}{(2+x)^2}e^x$ $\left[\textbf{Ans:} \frac{e^x}{2+x}+c\right]$

33. $e^x\sin x$ $\left[\textbf{Ans:} \frac{e^x}{2}(\sin x-\cos x)+c\right]$

34. Prove that $\int e^{ax}\cos bx\,dx=\dfrac{e^{ax}}{a^2+b^2}[a\cos bx+b\sin bx]+c$

35. Prove that

i. $\int \sqrt{a^2 + x^2}\, dx = \frac{x}{2}\sqrt{a^2 + x^2} + \frac{a^2}{2}\sinh^{-1}\frac{x}{a} + c$

ii. $\int \sqrt{a^2 - x^2}\, dx = \frac{x}{2}\sqrt{a^2 - x^2} + \frac{a^2}{2}\sinh^{-1}\frac{x}{a} + c$

iii. $\int \sqrt{x^2 - a^2}\, dx = \frac{x}{2}\sqrt{x^2 - a^2} + \frac{a^2}{2}\cosh^{-1}\frac{x}{a} + c$

Hint: $\int \sqrt{a^2 + x^2}\, dx = \int \sqrt{a^2 + x^2} \cdot 1dx$ and then integrat by parts.

2

Definite Integral and its Applications

2.0 Definite Integral

Let $f(x)$ be a continuous function defined over the closed interval $[a, b]$. Divide this interval into n sub-intervals by the points $a = x_0, x_1, x_2, \ldots\ldots x_{r-1}, x_r, \ldots x_n = b$ such that
$a = x_0 < x_1 < x_2 < \ldots\ldots x_{r-1} < x_r \ldots < x_n = b$

Let ξ be any point in the sub-interval (x_{r-1}, x_r)

Let $$S_n = \sum_{r=1}^{n} (x_n - x_{r-1}) f(\xi)$$

Then the limit of S_n *as* $n \to \infty$ in such a way that the length of each sub-interval $x_{r-1} - x_r \to 0$ is called the definite integral of $f(x)$ from and is denoted by $\int_a^b f(x)dx$

$$\text{Thus } \int_a^b f(x)dx = \lim_{n\to\infty} S_n = \lim_{n\to\infty} \sum_{r=1}^{n} (x_r - x_{r-1}) f(\xi)$$

Here are *a and b* called the limits of integration –in particular *a* is called the lower limit and the upper limt.

2.1 Fundamental Theorem of Integral Calculus

If $f(x)$ be a continuous in the interval $[a, b]$ and $\int f(x)dx = F(x) + c$

then $$\int_a^b f(x)dx = F(b) - F(a)$$

Proof of this theorem, is beyond the scope of the syllabus

For Example,

1. $\int_1^2 x^2 dx = \left.\frac{x^3}{3}\right]_1^2 = \frac{2^3}{3} - \frac{1^3}{3} = \frac{7}{3}$

2. $\int_0^{\pi/2} \cos x dx = \sin x]_0^{\pi/2} = \sin\frac{\pi}{2} - \sin 0 = 1$

3. $\int_1^3 (3x^2 - 2x + 1)dx = \left[\frac{3x^3}{3} - \frac{2x^2}{2} + x\right]_1^3 = \left[x^3 - x^2 + x\right]_1^3$

$$= \left(3^3 - 3^2 + 3\right) - (1 - 1 + 1) = 20$$

4. $\int_0^{\pi/4} \tan^2 x dx = \int_0^{\pi/4} (\sec^2 x - 1)dx = (\tan x - x)\Big]_0^{\pi/4}$

$$= \left(\tan\frac{\pi}{4} - \frac{\pi}{4}\right) - (\tan 0 - 0) = 1 - \frac{\pi}{4}$$

5. $\int_0^1 (2x-1)^8 dx = \left.\frac{(2x-1)^{8+1}}{(8+1)\cdot 2}\right]_0^1 = \left.\frac{(2x-1)^9}{18}\right]_0^1$

$$= \frac{1}{18}[(2-1)^9 - (0-1)^9] = \frac{1}{18}[1+1] = \frac{1}{9}$$

6. $\int\left(\sqrt{x} - \frac{1}{\sqrt{x}}\right)dx = \left.\frac{x^{3/2}}{3/2} + 2\sqrt{x}\right]_1^4 = \left.\frac{2x^{3/2}}{3} + 2\sqrt{x}\right]_1^4$

$$= \left[\frac{2}{3}\cdot 4^{4/3} + 2\sqrt{4}\right] - \left[\frac{2}{3} + 2\right] = \frac{16}{3} + 4 - \frac{2}{3} - 2 = \frac{20}{3}$$

WORKED EXAMPLES

Example 1 Evaluate $\int_0^{\pi/2} \sin^2 x\, dx$

Solution

Now

$$\int_0^{\pi/2} \sin^2 x\, dx = \int \frac{1 - \cos 2x}{2} dx$$

$$= \frac{1}{2}\left[\int 1dx - \int \cos 2x dx\right]$$

$$= \frac{1}{2}\left[x - \frac{\sin 2x}{2}\right] + c$$

$$\int_0^{\pi/2} \sin^2 x dx = \frac{1}{2}\left[x - \frac{\sin 2x}{2}\right]_0^{\pi/2}$$

$$= \frac{1}{2}\left[\frac{\pi}{2} - \frac{\sin \pi}{2}\right] - \frac{1}{2}\left[0 - \frac{\sin 0}{2}\right]$$

$$= \frac{\pi}{4} \qquad \text{since } \sin \pi = 0 = \sin 0$$

Example 2 Evaluate $\int_0^{\pi/2} \cos^2 x dx$

Solution

Now

$$\int \cos^2 x dx = \int \frac{1 + \cos 2x}{2} dx$$

$$= \frac{1}{2}\left[\int 1 dx + \int \cos 2x dx\right]$$

$$= \frac{1}{2}\left[x + \frac{\sin 2x}{2}\right] + c$$

$$\int_0^{\pi/2} \cos^2 x dx = \frac{1}{2}\left[x + \frac{\sin 2x}{2}\right]_0^{\pi/2}$$

$$= \frac{1}{2}\left[\frac{\pi}{2} + \frac{\sin \pi}{2}\right] - \frac{1}{2}\left[0 + \frac{\sin 0}{2}\right]$$

$$= \frac{\pi}{4} \text{ since } \sin \pi = 0 = \sin 0$$

Example 3 Evaluate $\int_0^{\pi/2} \cos^3 x dx$

Solution

We know that $\cos 3x = 4\cos^3 x - 3\cos x$

$$\therefore \quad 4\cos^3 x = \cos 3x + 3\cos x$$

$$\cos^3 x = \frac{1}{4}(\cos 3x + 3\cos x)$$

$$\int \cos^3 x dx = \frac{1}{4}\int (\cos 3x + 3\cos x) dx$$

$$= \frac{1}{4}\left[\int \cos 3x dx + \int 3\cos x dx\right] = \frac{1}{4}\left[\frac{\sin 3x}{3} + 3\sin x\right] + c$$

$$\int_0^{\pi/2} \cos^3 x dx = \frac{1}{4}\left[\frac{\sin 3x}{3} + 3\sin x\right]_0^{\pi/2}$$

$$= \frac{1}{4}\left[\frac{\sin\frac{3\pi}{2}}{3} + 3\sin\frac{\pi}{2}\right] - \frac{1}{4}\left[\frac{\sin 0}{3} + 3\sin 0\right]$$

$$= \frac{1}{4}\left(-\frac{1}{3} + 3\right) \quad \text{since } \sin\frac{3\pi}{2} = \sin\left(\pi + \frac{\pi}{2}\right) = -\sin\frac{\pi}{2} = -1$$

$$= \frac{1}{4}\cdot\frac{8}{3} = \frac{2}{2}$$

Example 4 Evaluate $\int_0^{\pi/2} \sin 5x \cos 3x dx$

Solution

We know that

$$\text{since } \sin A \cos B = \frac{1}{2}\{\sin(A+B) + \sin(A-B)\}$$

$$\sin 5x \cos 3x = \frac{1}{2}\{\sin(5x+3x) + \sin(5x-3x)\} = \frac{1}{2}(\sin 8x + \sin 2x)$$

$$\int_0^{\pi/2} \sin 5x \cos 3x dx = \frac{1}{2}\int_0^{\pi/2} (\sin 8x + \sin 2x) dx$$

$$= \frac{1}{2}\left(\frac{-\cos 8x}{8} - \frac{\cos 2x}{2}\right)\Bigg]_0^{\pi/2}$$

$$= \frac{1}{2}\left[\left(\frac{-\cos 4\pi}{8} - \frac{\cos \pi}{2}\right) - \frac{\cos 0}{8} - \frac{\cos 0}{2}\right)$$

$$= \frac{1}{2}\left[\left(-\frac{1}{8} + \frac{1}{2}\right) - \frac{1}{8} - \frac{1}{2}\right)$$

since $\cos 4\pi = \cos 0 = 1$ and $\cos \pi = -1$

$$= -\frac{1}{16} + \frac{1}{4} + \frac{1}{16} + \frac{1}{4} = \frac{1}{2}$$

Example 5 Evaluate $\int_1^e \log x dx$

Solution

On integrating by parts

$$\int \log x dx = \int \log x \cdot 1 \cdot dx$$

$$= \log x \cdot \int 1 \cdot dx - \int \left[\int 1 dx \right] \cdot \frac{d}{dx} (\log x) \cdot dx$$

$$= x \cdot \log x - \int x \cdot \frac{1}{x} \cdot dx$$

$$= (x \log x - x) + c$$

$$\int_0^e \log x dx = (x \log x - x) \Big]_1^e = (e \log e - e) - (0 - 1) = 1 \qquad [\because \quad \log e = 1]$$

Example 6 Evaluate $\int_0^{\pi/2} x \sin x dx$

Solution

On integrating by parts, we get,

$$\int x \sin x dx = x(-\cos x) - 1(-\sin x) + 0$$

$$= -x \cos x + \sin x + c$$

$$\int_0^{\pi/2} x \sin x dx = (-xcox + \sin x)]_0^{\pi/2}$$

$$= \left(-\frac{\pi}{2} \cos \frac{\pi}{2} + \sin \frac{\pi}{2} \right) - (0 + \sin 0) \quad = 1$$

since $\quad \sin \frac{\pi}{2} = 1 \;\; \cos \frac{\pi}{2} = 0$

Example 7 Evaluate $\int_0^{\pi/2} x^2 \cos x dx$

Solution

By using Bernoulli's formula, we get,

$$\int x^2 \cos x dx = x^2 (\sin x) - 2x(-\cos x) + 2(-\sin x) - 0$$

$$= x^2 \sin x + 2x \cos x - 2 \sin x + c$$

$$\int_0^{\pi/2} x^2 \cos x dx = (x^2 \sin x + 2x \cos x - 2 \sin x) \Big]_0^{\pi/2}$$

$$= \left[\frac{\pi^2}{2} \sin \frac{\pi}{2} + 2 \cdot \frac{\pi}{2} \cdot \cos \frac{\pi}{2} - 2 \sin \frac{\pi}{2} \right] - (0)$$

$$= \frac{\pi^2}{2} - 2 \quad \text{since } \sin \frac{\pi}{2} = 1, \;\; \cos \frac{\pi}{2} = 0$$

Example 8 Evaluate $\int_0^{\pi/2} x\sin^2 x dx$

Solution

Now

$$\int x\sin^2 x dx = \int x\cdot\frac{1}{2}(1-\cos 2x)dx$$

$$= \frac{1}{2}\int (x - x\cos 2x)dx$$

$$= \frac{1}{2}\left[\int xdx - \int x\cos 2xdx\right]$$

$$= \frac{1}{2}\left[\frac{x^2}{2} - \left\{x\cdot\frac{\sin 2x}{2} - 1\left(\frac{-\cos 2x}{4}\right)\right\}\right]$$

$$= \frac{1}{2}\left[\frac{x^2}{2} - \frac{x\sin 2x}{2} - \frac{\cos 2x}{4}\right] + c$$

$$\int_0^{\pi/2} x\sin^2 x dx = \frac{1}{2}\left[\frac{x^2}{2} - \frac{x\sin 2x}{2} - \frac{\cos 2x}{4}\right]_0^{\pi/2}$$

$$= \frac{1}{2}\left[\frac{\pi^2/4}{2} - \frac{\pi/2\cdot\sin\pi}{2} - \frac{\cos\pi}{4}\right] - \left[0 - 0 - \frac{\cos 0}{4}\right]$$

$$= \frac{1}{2}\left[\frac{\pi^2}{8} + \frac{1}{4}\right] + \frac{1}{2}\cdot\frac{1}{4} = \frac{\pi^2}{16} + \frac{1}{4}$$

2.2 Method of Substitution

When a substitution is used for evaluating the definite integral the limits of integration will change. Thus

$$\int_a^b f(x)dx = \int_a^b f\{\phi(t)\},\phi'(t)\cdot dt$$

Where the substitution is $\phi(t)$

Also $t = t_1$ when $x = a$ and $t = t_2$ when $x = b$

The following examples illustrate the procedure.

Example 9 Evaluate $\int_0^{\pi/2} \cos^3 x\sin xdx$

Solution

substitution $\cos x = t$, so that $-\sin xdx = dt$ or $\sin xdx = dt$

when $x = 0$, $t = \cos 0 = 1$

when $x=\frac{\pi}{2}$, $t=\cos\frac{\pi}{2}=0$

Given integral

$$\int_0^{\pi/2}\cos^3 x\sin x dx = \int_1^0 t^3(-dt) = -\int_1^0 t^3 dt$$

$$= -\frac{t^4}{4}\Big]_1^0 = -\left[\frac{0}{4}-\frac{1}{4}\right] = \frac{1}{4}$$

Example 10 Evaluate $\int_0^{\pi/4}\tan^4 x\sec^2 x dx$

Solution

Substitution $\tan x = t$, so that $\sec^2 x dx = dt$

when $x = 0$, $t=\tan 0 = 0$

when $x = \frac{\pi}{4}$, $t=\tan\frac{\pi}{4}=1$

Given integral

$$\int_0^{\pi/4}\tan^4 \sec^2 x dx = \int_0^1 t^4 dt$$

$$= \int_0^1 t^4 dt = \frac{t^5}{5}\Big]_0^1 = \left[\frac{1}{5}-\frac{0}{5}\right] = \frac{1}{5}$$

Example 11 Evaluate $\int_0^1 (x^2+2x+3)^2(x+1)dx$

Solution

Substitution $x^2+2x+3=t$

Differentiating, $(2x+2)dx = dt$, $2(x+1)dx = dt$ or $(x+1)dx=\frac{dt}{2}$

when $x = 0$, $t = 0+0+3 = 3$

when $x = 0$, $t = 1+2+3 = 6$

Given integral

$$\int_0^1 (x^2+2x+3)^2(x+1)dx = \int_3^6 t^2\frac{dt}{2} = \frac{1}{2}\int_3^6 t^2 dt$$

$$= \frac{t}{6}\Big]_3^6 = \frac{1}{6}[6^3-3^3] = \frac{1}{6}[216-27] = \frac{189}{6} = \frac{63}{2}$$

Example 12 Evaluate $\int_0^1 \frac{(\tan^{-1} x)^2}{1+x^2} dx$

Solution

Substitution $\tan^{-1} x = t$, so that $\frac{1}{1+x^2} dx = dt$

when $x = 0, \; t = \tan^{-1} 0 = 0$

when $x = 1, t = \tan^{-1} 1 = \frac{\pi}{4}$

Given integral

$$\int_0^1 \frac{(\tan^{-1} x)^2}{1+x^2} dx = \int_0^{\pi/4} t^2 dt$$

$$= \left.\frac{t^3}{3}\right]_0^{\frac{\pi}{4}} = \frac{1}{3}\left[\left(\frac{\pi}{4}\right)^3 - 0\right] = \frac{\pi^3}{192}$$

Example 13 Evaluate $\int_0^1 \frac{(\sin^{-1} x)^3}{\sqrt{1-x^2}} dx$

Solution

Substitution $\sin^{-1} x = t$, so that $\frac{1}{\sqrt{1-x^2}} dx = dt$

when $x = 0, \; t = \sin^{-1} 0 = 0$

when $x = 1, t = \sin^{-1} 1 = \frac{\pi}{2}$

Given integral

$$\int_0^1 \frac{(\sin^{-1} x)^3}{\sqrt{1-x^2}} dx = \int_0^{\pi/2} t^3 dt$$

$$= \left.\frac{t^4}{4}\right]_0^{\pi/2} = \frac{1}{4}\left[\left(\frac{\pi}{2}\right)^4 - 0\right] = \frac{\pi^4}{64}$$

Example 14 Evaluate $\int_0^1 \frac{\sin x}{1+\cos^2 x} dx$

Solution

Substitution $\cos x = t$, so that $-\sin x dx = dt$ or $\sin x dx = -dt$

when $\quad x = \frac{\pi}{2}, \quad t = \cos\frac{\pi}{2} = 0$

when $\quad x = 0, \quad t = \cos 0 = 1$

Given integral

$$\int_0^{\pi/2} \frac{\sin x}{1+\cos^2 x}\, dx = \int_1^0 \frac{-dt}{1+t^2} = -\tan^{-1} x \Big]_1^0 = -\left[\tan^{-1} 0 - \tan^{-1} 1\right]$$

$$= -\left[0 - \frac{\pi}{4}\right] = \frac{\pi}{4}$$

Example 15 Evaluate $\int_0^a \sqrt{a^2 - x^2}\, dx$

Solution

Substitution $x = \sin\theta$, Hence $dx = a\cos\theta d\theta$

Now $\quad \sin\theta = \frac{x}{a} \quad$ or $\quad \theta = \sin^{-1}\frac{x}{a}$

when $\quad x = 0, \quad \theta = \sin^{-1}\frac{0}{a} = \sin^{-1} 0 = 0$

when $\quad x = a, \quad \theta = \sin^{-1}\frac{a}{a} = \sin^{-1} 1 = \frac{\pi}{2}$

Given integral

$$\int_0^a \sqrt{a^2 - x^2}\, dx = \int_0^{\pi/2} \sqrt{a^2 - a^2 \sin^2\theta} \cdot a\cos\theta d\theta$$

$$= \int_0^{\pi/2} \sqrt{a^2(1 - \sin^2\theta)} \cdot a\cos\theta d\theta$$

$$= \int_0^{\pi/2} \sqrt{a^2\cos^2\theta} \cdot a\cos\theta d\theta$$

$$= \int_0^{\pi/2} a^2\cos^2\theta d\theta$$

$$= a^2 \int_0^{\pi/2} \frac{1}{2}(1 + \cos 2\theta) d\theta$$

$$= \frac{a^2}{2} \int_0^{\pi/2} (1 + \cos 2\theta) d\theta$$

$$= \frac{a^2}{2}\left(\theta + \frac{\sin 2\theta}{2}\right)\Big]_0^{\pi/2}$$

$$= \frac{a^2}{2}\left[\left(\frac{\pi}{2} + \frac{\sin \pi}{2}\right) - \left(0 + \frac{\sin 0}{2}\right)\right]$$

$$= \frac{\pi a^2}{4} \quad \text{since } \sin \pi = 0$$

EXERCISES

1. $\int_1^2 (x^2 - 1)dx$ $\left[\textbf{Ans:} \frac{4}{3}\right]$

2. $\int_{-1}^1 (x^2 - 5x + 2)dx$ $\left[\textbf{Ans:} \frac{14}{3}\right]$

3. $\int_0^1 (x^3 - 3x^2 + 2x + 1)dx$ $\left[\textbf{Ans:} \frac{5}{4}\right]$

4. $\int_1^3 \left(x + \frac{1}{x}\right)^2 dx$ $\left[\textbf{Ans:} \frac{40}{3}\right]$

5. $\int_0^1 (2x + 3x)^2 dx$ $\left[\textbf{Ans:} \frac{49}{3}\right]$

6. $\int_0^1 (3x - 1)^3 dx$ $\left[\textbf{Ans:} \frac{5}{4}\right]$

7. $\int_0^1 (x^2 + 1)^2 dx$ $\left[\textbf{Ans:} \frac{28}{15}\right]$

8. $\int_0^3 \frac{dx}{\sqrt{9 - x^2}}$ $\left[\textbf{Ans:} \frac{\pi}{2}\right]$

9. $\int_0^{\pi/4} \sin 2x dx$ $\left[\textbf{Ans:} \frac{1}{2}\right]$

10. $\int_0^{\pi/4} \sec^2 3x dx$ $\left[\textbf{Ans:} -\frac{1}{3}\right]$

11. $\int_0^{\infty} e^{-xdx}$ **[Ans :1]**

12. $\int_0^{3} \frac{dx}{x^2+9}$ $\left[\textbf{Ans}: \frac{\pi}{12}\right]$

13. $\int_6^{8} \frac{dx}{x^2-16}$ $\left[\textbf{Ans}: \frac{1}{8}\log\frac{5}{3}\right]$

14. $\int_0^{4} \frac{dx}{25-x^2}$ $\left[\textbf{Ans}: \frac{1}{10}\log 6\right]$

15. $\int_0^{\pi/2} \cos^2 x dx$ $\left[\textbf{Ans}: \frac{\pi}{4}\right]$

16. $\int_0^{\pi/2} x \cos x \, dx$ $\left[\textbf{Ans}: \frac{\pi}{2}-1\right]$

17. $\int_0^{\pi/2} x \sin 3x dx$ $\left[\textbf{Ans}: -\frac{1}{9}\right]$

18. $\int_0^{\pi/2} \sin 3x \cos x dx$ $\left[\textbf{Ans}: \frac{1}{2}\right]$

19. $\int_0^{\pi/2} \sin 3x \sin x dx$ **[Ans : 0]**

20. $\int_0^{\pi/4} \cos 4x \cos 2x dx$ $\left[\textbf{Ans}: \frac{1}{6}\right]$

21. $\int_0^{\pi/2} x^2 \sin x dx$ $[\textbf{Ans}: \pi - 2]$

22. $\int_0^{\pi/2} \sin^3 x dx$ $\left[\textbf{Ans}: \frac{2}{3}\right]$

23. $\int_1^{2} \log x dx$ $[\textbf{Ans}: 2\log 2 - 1]$

2.3 Properties of the Definite Integral

1. $\int_a^b f(x)dx = \int_a^b f(t)dt$

In the L.H.S substitute $x = t \;\therefore\; dx = dt$ *Also when* $x = a,\ t = a$ when $x = b,\ t = b$

$$\therefore \quad \int_a^b f(x)dx = \int_a^b f(t)dt$$

Thus the value of the definite integral is independent of the variable.

2. $\int_a^b f(x)dx = -\int_b^a f(x)dt$

$$\text{Let} \quad \int_a^b f(x)dx = F(x)+c$$

$$\therefore \quad \int f(x)dx = F(b)-F(a) = -[F(a)-F(a) = -\int_b^a f(x)dx$$

3. $\int_a^b f(x)dx = \int_a^c f(x)dx + \int_c^b f(x)dx$, where $a < c < b$

$$\text{Let} \quad \int_a^b f(x)dx = F(x)+c$$

Hence

$$\int_a^b f(x)dx = F(b)-F(a)$$

$$= [F(c)-F(a)]+[F(b)-F(c)]$$

$$= \int_a^c f(x)dx + \int_c^b f(x)dx$$

4. $\int_0^a f(x)dx = \int_0^a f(a-x)dx$

In the L.H.S substitute $x = a - t$ or $a - x = t$ *so that* $dx = -dt$
Also when $x = 0,\ t = a$ *when* $x = a,\ t = 0$

$$\therefore \quad \int_0^a f(x)dx = \int_a^0 f(a-t)(-dt)$$

$$= -\int_a^0 f(a-t)dt$$

$$= \int_0^a f(a-t)dt \quad \textit{by property 2}$$

$$= \int_a^0 f(a-x)dt \quad \textit{by property 1}$$

As an important special case of this property, we have

$$\int_0^{\pi/2} f(\sin x)dx = \int_0^{\pi/2} f\left\{\sin\left(\frac{\pi}{2}-x\right)\right\}dx$$

$$= \int_0^{\pi/2} f(\cos x)dx$$

5. $$\int_a^b f(x)dx = \int_a^b f(a+b-x)dx$$

In the L.H.S substitute $a + b - x = t$ *or* $x = a + b + t$ *so that* $dx = -dt$
Also when $x = a$, $t = b$ *when* $x = b$, $t = a$

$$\therefore \quad \int_a^b f(x)dx = \int_b^a (f(a+b-t)(-dt)$$

$$= \int_b^a f(a+b-t)dt \quad \textit{by property 2}$$

$$= \int_a^b f(a+b-x)dt \quad \textit{by property 1}$$

6. $$\int_0^a f(x)dx = \int_0^{a/2} f(x)dx + \int_0^{a/2} f(a-x)dx$$

Now $$\int_0^a f(x)dx = \int_0^{a/2} f(x)dx + \int_{a/2}^a f(x)dx$$

by property 3 since $0 < \frac{a}{2} < a$

In the second integral substitute $x = a - t$ so that $dx = -dt$

Also when $t = a - x$ *when* $x = \frac{a}{2}, t = \frac{a}{2}$ *when* $x = a, t = 0$

$$\int_0^a f(x)dx = \int_0^{a/2} f(x)dx + \int_{a/2}^0 f(a-t)(-dt)$$

$$= \int_0^{a/2} f(x)dx - \int_{a/2}^0 f(a-t)dt$$

$$= \int_0^{a/2} f(x)dx + \int_0^{a/2} f(a-t)dt$$

$$= \int_0^{\frac{a}{2}} f(x)dx + \int_0^{\frac{a}{2}} f(a-x)dx \text{ by using properties 1 and 2}$$

Corollary : $$\int_0^a f(x)dx = \begin{cases} 2\int_0^{a/2} f(x)dx, \text{ if } f(a-x) \text{ is odd} \\ 0 \quad ,\text{if } f(a-x) = -f(x) \end{cases}$$

7. $$\int_{-a}^a f(x)dx = \begin{cases} 2\int_0^{a/2} f(x)dx, \text{ if } f(a-x) \text{ is even} \\ 0 \quad ,\text{if } f(x) \text{is odd} \end{cases}$$

$$\int_{-a}^a f(x)dx = \int_{-a}^a f(x)dx + \int_0^a f(x)dx$$

by property 3 since $-a < 0 < a$

In the first integral substitute $x = -t$ *so that* $dx = -dt$

The limits corresponding to t are a and 0.

$$\therefore \int_{-a}^a f(x)dx = \int_a^0 f(-t)(-dt) + \int_0^a f(x)dx = -\int_a^0 f(-t)dt + \int_0^a f(x)dx$$

$$= \int_0^a f(-x)dt + \int_0^a f(x)dx \text{ by using properties 1 and 2}$$

$$= \int_0^a f(-x) + f(x)dx$$

$$= \begin{cases} 2\int_0^a f(x)dx, \text{ if } & f(-x) = f(x) \text{ i.e., } f(x) \text{is even} \\ 0 & , \text{if } \quad f(-x) = -f(x) \text{ i.e.,} f(x) \text{is odd} \end{cases}$$

WORKED EXAMPLES

Example 1 Prove that $\int_0^{\pi/2} \sin^2 x dx = \int_0^{\pi/2} \cos^2 x dx = \frac{\pi}{4}$

Solution

Let

$$I = \int_0^{\pi/2} \sin^2 x dx \qquad \dots 1$$

$$I = \int_0^{\pi/2} \sin^2\left(\frac{\pi}{2} - x\right) dx \quad \textit{by property 4}$$

$$= \int_0^{\pi/2} \cos^2 x dx \qquad \dots 2$$

Adding (1) and (2)

$$2I = \int_0^{\pi/2} \sin^2 x dx + \int_0^{\pi/2} \cos^2 x dx$$

$$= \int_0^{\pi/2} (\sin^2 x + \cos^2 x) dx = \int_0^{\pi/2} 1 \cdot dx = x\Big]_0^{\pi/2} = \frac{\pi}{2}$$

$$\therefore \quad I = \frac{\pi}{4}$$

Example 2 Evaluate $\int_0^{\pi/2} \frac{\sqrt{\sin x}}{\sqrt{\sin x} + \sqrt{\cos x}} dx$

Solution

Let

$$\therefore \quad I = \int_0^{\pi/2} \frac{\sqrt{\sin x}}{\sqrt{\sin x} + \sqrt{\cos}} dx \qquad \dots 1$$

$$I = \int_0^{\pi/2} \frac{\sqrt{\sin\left(\frac{\pi}{2}-x\right)}}{\sqrt{\sin\left(\frac{\pi}{2}-x\right)}+\sqrt{\cos\left(\frac{\pi}{2}-x\right)}}\,dx, \text{ by property 4,}$$

$$I = \int_0^{\pi/2} \frac{\sqrt{\cos x}}{\sqrt{\cos x}+\sqrt{\sin x}}\,dx \qquad \ldots 2$$

Adding (1) and (2)

$$2I = \int_0^{\pi/2} \frac{\sqrt{\sin x}+\sqrt{\cos x}}{\sqrt{\cos x}+\sqrt{\sin x}}\,dx$$

$$= \int_0^{\pi/2} 1\cdot dx = x\Big]_0^{\pi/2} = \frac{\pi}{2}$$

$$\therefore \quad I = \frac{\pi}{4}$$

Example 3 Prove that $\int_0^{\pi} \frac{x\sin x}{1+\cos^2 x}dx = \frac{\pi^2}{4}$

Solution

Let

$$I = \int_0^{\pi} \frac{x\sin x}{1+\cos^2 x}dx$$

$$= \int_0^{\pi} \frac{(\pi-x)\sin(\pi-x)}{1+\cos^2(\pi-x)}dx$$

$$= \int_0^{\pi} \frac{(\pi-x)\sin x}{1+\cos^2 x}dx \qquad \text{by property 4}$$

$$= \pi\int_0^{\pi} \frac{\sin x}{1+\cos^2 x} - \int_0^{\pi} \frac{x\sin x}{1+\cos^2 x}dx$$

$$I = \pi\int_0^{\pi} \frac{\sin x}{1+\cos^2 x}dx - I$$

$$2I = \pi\int_0^{\pi} \frac{\sin x}{1+\cos^2 x}dx$$

Now put $\cos x = t \quad \therefore \quad -\sin x = dt$ *i.e.,* $\sin x\, dx = -dx$

when $x = 0, \; t = \cos 0 = 1$

when $x = \pi, \; t = \cos\pi = -1$

$$2I = \pi \int_{1}^{-1} \frac{-dt}{1+t^2}$$

$$= \pi \int_{-1}^{1} \frac{-dt}{1+t^2}$$

$$= \pi \tan^{-1} t \Big]_{-1}^{1}$$

$$= \pi\left[\tan^{-1} 1 - \tan^{-1}(-1)\right]$$

$$= \pi\left[\frac{\pi}{4} + \frac{\pi}{4}\right]$$

$$2I = \pi\left[\frac{\pi}{2}\right] = \frac{\pi^2}{2} \quad \therefore \; I = \frac{\pi^2}{4}$$

Example 4 Prove that $\int_{0}^{\pi/4} \log(1+\tan x)dx = \frac{\pi}{8}\log 2$

Solution

Let

$$I = \int_{0}^{\pi/4} \log(1+\tan x)dx$$

$$= \int_{0}^{\pi/4} \log\left[1+\tan\left(\frac{\pi}{4}-x\right)\right]dx \quad \textit{by property 4}$$

$$= \int_{0}^{\pi/4} \log\left[1+\frac{1-\tan x}{1+\tan x}\right]dx$$

$$= \int_{0}^{\pi/4} \log\left[\frac{2}{1+\tan x}\right]dx$$

$$I = \int_{0}^{\pi/4} \log 2 dx - \int_{0}^{\pi/2} \log(1+\tan x)dx$$

$$= \log 2 \cdot \int_{0}^{\pi/4} 1 dx - I$$

$$2I = \log 2 \cdot x\Big]_0^{\pi/4}$$

$$= \frac{\pi}{4}\log 2$$

$$\therefore \qquad I = \frac{\pi}{8}\log 2$$

Example 5 Evaluate $\int_{\pi/6}^{\pi/3} \frac{1}{1+\sqrt{\tan x}}dx$

Solution

Let

$$I = \int_{\pi/6}^{\pi/3} \frac{1}{1+\sqrt{\tan x}}dx \qquad \ldots 1$$

$$= \int_{\pi/6}^{\pi/3} \frac{1}{1+\sqrt{\tan\left(\frac{\pi}{6}+\frac{\pi}{3}-x\right)}}dx \quad \textit{by property 5}$$

$$= \int_{\pi/6}^{\pi/3} \frac{1}{1+\sqrt{\tan\left(\frac{\pi}{2}-x\right)}}dx$$

$$= \int_{\pi/6}^{\pi/3} \frac{1}{1+\sqrt{\cot x}}dx$$

$$= \int_{\pi/6}^{\pi/3} \frac{\sqrt{\tan x}}{1+\sqrt{\tan x}}dx \qquad \ldots 2$$

Adding (1) *and* (2)

$$2I = \int_{\pi/6}^{\pi/3} \frac{1+\sqrt{\tan x}}{1+\sqrt{\tan x}}dx$$

$$= \int_{\pi/6}^{\pi/3} 1 \cdot dx = \Big]_{\pi/6}^{\pi/3}$$

$$= \frac{\pi}{3} - \frac{\pi}{6} = \frac{\pi}{6}$$

$$I = \frac{\pi}{12}$$

Example 6 Prove that $\int_0^{\pi/2} \log \sin x dx = \frac{\pi}{2} \log \frac{1}{2}$

Solution

$$I = \int_0^{\pi/2} \log \sin x dx \quad \text{... (1)}$$

$$= \int_0^{\pi/2} \log \sin\left(\frac{\pi}{2} - x\right) dx \quad \textit{by the property 4,}$$

$$= \int_0^{\pi/2} \log \cos x dx \quad \text{... (2)}$$

Adding (1) and (2)

$$2I = \int_0^{\pi/2} (\log \sin x + \log \cos x) dx$$

$$= \int_0^{\pi/2} (\log \sin x \cos x) dx$$

$$= \int_0^{\pi/2} \log \frac{\sin 2x}{2} dx, \quad \text{where } \sin x \cos x = \frac{\sin 2x}{2}$$

$$= \int_0^{\pi/2} [\log \sin 2x - \log 2] dx$$

$$= \int_0^{\pi/2} \log \sin 2x - \int_0^{\pi/2} \log 2 dx$$

Now substitute $2x = t$ *in the first integral on RHS so that* $dx = \frac{dt}{2}$

If $x = 0,\ t = 0$, *and if* $x = \frac{\pi}{2},\ t = \pi$

$$2I = \int_0^{\pi} \log \sin t \frac{dt}{2} - \frac{\pi}{2} \log 2$$

$$= \frac{1}{2} \cdot 2 \int_0^{\pi} \log \sin t dt - \frac{\pi}{2} \log 2 = I - \frac{\pi}{2} \log \frac{1}{2}$$

$$\therefore \quad I = \frac{\pi}{2} \log \frac{1}{2}$$

EXERCISES

Evaluate the following

1. $\int_0^{\pi/2} \frac{\sin x}{\sin x + \cos x} dx$ $\left[\textbf{Ans}: \frac{\pi}{4}\right]$

2. $\int_0^{\pi/2} \frac{\sqrt{\tan x}}{\sqrt{\tan x} + \sqrt{\cot x}} dx$ $\left[\textbf{Ans}: \frac{\pi}{4}\right]$

3. $\int_0^{\pi/2} \frac{\sqrt[3]{\cos x}}{\sqrt[3]{\sin x} + \sqrt[3]{\cos x}} dx$ $\left[\textbf{Ans}: \frac{\pi}{4}\right]$

4. $\int_0^{\pi/2} \frac{\tan x}{\tan x + \cot x} dx$ $\left[\textbf{Ans}: \frac{\pi}{4}\right]$

5. $\int_0^{\pi/2} \frac{1}{1 + \tan x} dx$ $\left[\textbf{Ans}: \frac{\pi}{4}\right]$

6. $\int_0^{\pi/2} \frac{1}{1 + \sqrt{\cot x}} dx$ $\left[\textbf{Ans}: \frac{\pi}{4}\right]$

7. $\int_0^{2} \frac{\sqrt{x+3}}{\sqrt{x+3} + \sqrt{5-x}} dx$ [**Ans** : 1]

8. $\int_0^{\pi/2} \frac{\sin x - \cos x}{1 + \sin x \cdot \cos x}$ [**Ans** : 0]

9. $\int_0^{a} \frac{1}{x + \sqrt{a^2 - x^2}} dx$ [Hint: put $x = \sin \theta$] $\left[\textbf{Ans}: \frac{\pi}{4}\right]$

10. $\int_0^{\infty} \frac{x}{(1+x) \cdot (1+x^2)} dx$ [Hint: put $x = \tan \theta$] $\left[\textbf{Ans}: \frac{\pi}{4}\right]$

11. $\int_0^{\pi/2} \log \tan x dx$ [**Ans** : 0]

12. $\int_0^{1} \frac{\log(1+x)}{1+x^2} dx$ $\left[\textbf{Ans}: \frac{\pi}{8} \log 2\right]$

13. $\int_0^{\pi} x\cos^2 x dx$ $\left[\textbf{Ans}: \frac{\pi^2}{4}\right]$

14. $\int_0^{\pi} \frac{x}{1+\sin x} dx$ [**Ans** : π]

15. $\int_0^{1} x(1-x)^n dx$ $\left[\textbf{Ans}: \frac{1}{(n+1)(n+2)}\right]$

16. $\int_0^{a} x\sqrt{a-x} dx$ $\left[\textbf{Ans}: \frac{4}{15}a^{\frac{8}{2}}\right]$

17. $\int_1^{3} \frac{\sqrt{x}}{\sqrt{4-x}+\sqrt{x}} dx$ [**Ans** : 1]

18. $\int_0^{\pi} \frac{x\sin x}{1+\sin x} dx$ $\left[\textbf{Ans}: \pi\left(\frac{\pi}{2}-1\right)\right]$

19. $\int_0^{\pi} \frac{x\tan x}{\sec x+\cos x} dx$ $\left[\textbf{Ans}: \frac{\pi^2}{4}\right]$

20. $\int_0^{\pi/2} \frac{\sin^2 x}{\sin x+\cos x} dx$ $\left[\textbf{Ans}: \frac{1}{\sqrt{2}}\log\left(\sqrt{2+1}\right)\right]$

21. $\int_a^{\pi} \frac{xdx}{a^2\cos^2 x+b^2\sin^2 x}$ $\left[\textbf{Ans}: \frac{\pi^2}{8ab}\right]$

22. $\int_a^{\frac{\pi}{2}} \frac{\sin^2 x}{1+\sin\cos x} dx$ $\left[\textbf{Ans}: \frac{\pi}{2\sqrt{3}}\right]$

2.4 Define Integral as the Limit of a Sum

We know that,

$$\int_a^b f(x)dx = \lim_{n\to\infty} \sum_{r=1}^{n} f(\xi_r)(x_r - x_{r-1})$$

where $\xi_r \in [x_{r-1}, x_r]$

We take all the sub-intervals of equal length say h and ξ_r to coincide with Then

$$\int_a^b f(x)dx = \lim_{n\to\infty}\sum_{r=1}^{n} f(x_{r-1})h$$

Clearly $x_{r-1} = a + (r-1)h$, where $r = 1, 2, \ldots n$

$$\int_a^b f(x)dx = \lim_{n\to\infty} h\sum_{r=1}^{n} f(a+(r-1)h$$

In particulars if $a = 0$ and $b = 1$, we have $rh = 1$ and $h = \frac{1}{n}$

$$\text{Hence } \int_0^1 f(x)dx = \lim_{n\to\infty}\frac{1}{n} f\left(\frac{r-1}{n}\right)$$

$$\int_0^1 f(x)dx = \lim_{n\to\infty}\frac{1}{n}\sum_{r=0}^{n-1} f\left(\frac{r}{n}\right) \quad \text{on replacing } r-1 \text{ by } r$$

Example 1 Using the definite integral as the limit of a sum, evaluate

$$\lim_{n\to\infty}\sum_{r=0}^{n-1}\frac{n}{n^2+r^2}$$

Solution

$$\text{Required limit} = \lim_{n\to\infty}\sum_{r=0}^{n-1}\frac{n}{n^2+r^2} = \lim_{n\to\infty}\frac{1}{n}\sum_{r=0}^{n-1}\frac{n^2}{n^2+r^2}$$

$$= \lim_{n\to\infty}\sum_{r=0}^{n-1}\frac{1}{1+\left(\frac{r^2}{n^2}\right)} = \lim_{n\to\infty}\frac{1}{n}\sum_{r=0}^{n-1}\frac{1}{1+\left(\frac{r}{n}\right)^2}$$

$$= \int_0^1 \frac{1}{1+x^2}dx = \tan^{-1}x\Big]_0^1$$

$$= \tan^{-1}1 - \tan^{-1}0 = \frac{\pi}{4} - 0 = \frac{\pi}{4}$$

Example 2 Evaluate $\lim_{n\to\infty}\left[\frac{1}{n+1}+\frac{1}{n+2}+\ldots\frac{1}{n+n}\right]$

Solution

$$\text{Required limit} = \lim_{n\to\infty}\sum_{r=1}^{n}\frac{1}{n+r} = \lim_{n\to\infty}\frac{1}{n}\sum_{r=1}^{n}\frac{1}{1+(r/n)}$$

$$= \int_0^1 \frac{1}{1+x}dx = \log(1+x)\Big]_0^1 = \log 2$$

Example 3 Evaluate $\lim\limits_{n\to\infty}\left[\frac{1}{4n^2-1^2}+\frac{1}{\sqrt{4n^2-2^2}}+.....+\frac{1}{\sqrt{4n^2-n^2}}\right]$

Solution

$$\text{Required limit} = \lim_{n\to\infty}\sum_{r=1}^{n}\frac{1}{\sqrt{4n^2-r^2}}$$

$$= \lim_{n\to\infty}\frac{1}{n}\sum_{r=1}^{n}\frac{n}{\sqrt{4n^2-r^2}}$$

$$= \lim_{n\to\infty}\frac{1}{n}\sum_{r=1}^{n}\frac{1}{\sqrt{4-\frac{r^2}{n^2}}} = \lim_{n\to\infty}\frac{1}{n}\sum_{r=1}^{n}\frac{1}{\sqrt{4-\left(\frac{r}{n}\right)^2}}$$

$$= \int_0^1\frac{1}{\sqrt{4-x^2}}dx = \sin^{-1}\frac{x}{2}\Bigg]_0^1$$

$$= \sin^{-1}\frac{1}{2}-\sin^{-1}0 = \frac{\pi}{6}$$

EXERCISES

Using the definite integral as the limits of a sum

1. Evaluate $\lim\limits_{n\to\infty}\left[\frac{1}{\sqrt{n^2-0}}+\frac{1}{\sqrt{n^2-1^2}}+...+\frac{1}{\sqrt{n^2-(n-1)^2}}\right]$

   $\left[\text{Ans:}\frac{\pi}{2}\right]$

2. Evaluate $\lim\limits_{n\to\infty}\left[\frac{n}{(n+1)^2}+\frac{1}{(n+2)^2}+\frac{n}{(n+3)^3}...+\frac{n}{(n+n)^2}\right]$ $\left[\text{Ans:}\frac{1}{2}\right]$

3. Evaluate $\lim\limits_{n\to\infty}\left[\frac{1}{3n+1}+\frac{1}{3n+2}+....+\frac{1}{3n+n}\right]$ $\left[\text{Ans:}\log\frac{4}{3}\right]$

4. Evaluate

$$\lim_{n\to\infty} n^2\left[\frac{1}{(n^2+1^2)^{3/2}}+\frac{1}{(n^2+2^2)^{3/2}}+...+\frac{1}{(n^2+(n-1)^2)^{3/2}}\right]$$

   $\left[\text{Ans:}\frac{1}{\sqrt{2}}\right]$

5. $\lim\limits_{n\to\infty} n^2\left[\dfrac{n^2}{(n+1)^3}+\dfrac{n^2}{(n+2)^3}+\ldots+\dfrac{n^2}{(n+n)^3}\right]$ $\left[\text{Ans:}\dfrac{3}{8}\right]$

6. $\lim\limits_{n\to\infty}\left[\dfrac{1}{n^2+1^2}+\dfrac{1}{n^2+2^2}+\dfrac{1}{n^2+3^2}\ldots+\dfrac{1}{n^2+n^2}\right]$ $\left[\text{Ans:}\dfrac{1}{2}\log 2\right]$

2.5 Application of the Definite Integral to Areas

2.5.1 Geometrical meaning of the definite integral

Let $f(x)$ be a continuous function defined in the closed interval $[a, b]$. Then we have defined the definite integral $\int_a^b f(x)dx$ as,

$$\int_a^b f(x)dx = \lim_{n\to\infty}\sum_r^n f(\xi_r)(x_r - x_{r-1})$$

Where $\xi_r \in [x_r - x_{r-1}]$ and the limit is taken in such a way that the length of each sub-interval $x_r - x_{r-1} \to 0$

In the above definition we can take $\xi_r = x_{r-1}$ or $\xi_r = x_r$ where $r = 1, 2, \ldots. n$

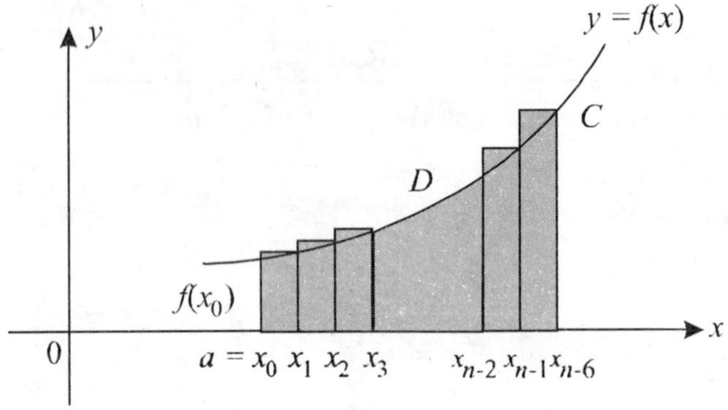

Fig. 2.1

Let $s_n = \sum_{r=1}^{n} f(x_r)\Delta_r$ and $s_n = \sum_{r=1}^{n} f(x_r - 1)\Delta_r$ where $\Delta_r = x_r - x_r - 1$

From the above figure it is clear that

$$S_n = \Delta_1 f(x_1) + \Delta_2 f(x_2) + \ldots + \Delta_n f(x_n)$$
$$= \text{sum of the areas of the outer rectangles.}$$

and

$$S_n = \Delta_1 f(x_0) + \Delta_2 f(x_1) + \ldots + \Delta_n f(x_{n-1})$$
$$= \text{sum of the areas of the inner rectangles.}$$

Hence if A is the area bounded by the curve $y = f(x)$, the x = axis and the ordinates at $x = a$ and $x = b$ then

$S_n < A < S_n$ for all n.

Now $\quad \lim_{n\to\infty} S_n = \lim_{n\to\infty} S_n = \int_a^b f(x)dx$

As $n \to \infty$, in such a way that the length of each partial sub-interval $\Delta_r \to 0$ both $\lim_{n\to\infty} S_n$ and $\lim_{n\to\infty} S_n$ exist and coincide with A.

Therefore, the definite integral $\int_a^b f(x)dx$ or $\int_a^b y\,dx$ represents area bounded by the curve $y = f(x)$, the x – axis and the ordinate at $x = a$ and $x = b$ i.e., the area $ABCDE$ as shown in the figure 4.1.

Note:

1. When the curve is below the x-axis $\int_a^b f(x)dx$ is negative. In such cases we take the absolute value of the integral.
2. If the curve $y = f(x)$ crosses the x-axis at c, where $a < c < b$ then

$$\int_a^b f(x)dx = \left|\int_a^c f(x)dx\right| + \left|\int_c^b f(x)dx\right|$$

3. The definite integral $\int_a^b x\,dy$ represents the area bounded by the curve $x = g(y)$, the y - axis and the abscissa $y = a$ and $y = b$.

2.5.2 Area between two curves

Suppose $y = f(x)$ and $y = g(x)$ are two continuous functions defined in the interval $[a, b]$.

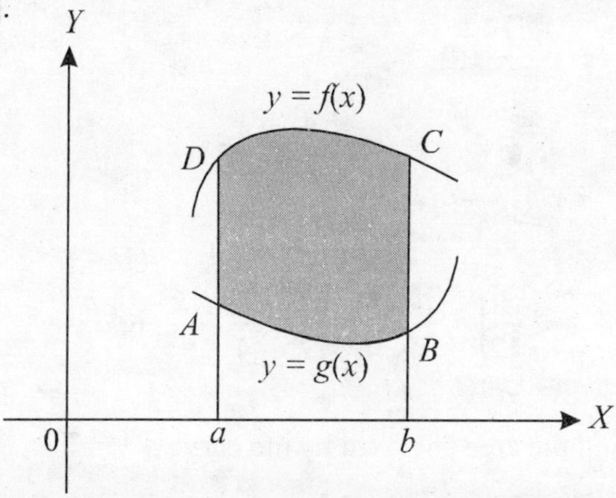

Fig. 2.2

If $f(x) \geq g(x)$ for all $x = \in [a,b]$ then area between two curves $y = f(x)$ and $y = g(x)$ and the ordinates $x = a$ and $x = b$ is given by

$$\int_a^b f(x)dx - \int_a^b g(x)dx = \int_a^b \{f(x) - g(x)\}\,dx$$

This formula holds good even if the curve $y = f(x)$ is below the x - axis

WORKED EXAMPLES

Examples 1 Find the area bounded by the curve $x^2 - 4y$ the x-axis and the ordinates $x = 1$ and $x = 4$

Solution

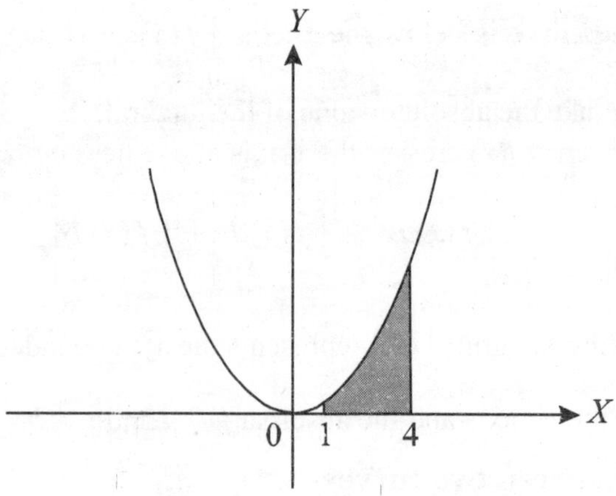

Fig. 2.3

$$\text{Required area} = \int_1^4 y \cdot dx$$

$$= \int_1^4 \frac{x^2}{4} \cdot dx = \frac{1}{4} \cdot \frac{x^3}{3} = \left.\frac{x^3}{12}\right]_1^4$$

$$= \frac{1}{12}\left[4^3 - 1^3\right] = \frac{63}{12} = \frac{21}{4} \text{ sq. units}$$

Example 2 Find the area enclosed by the curve $y = \dfrac{x^2}{x^2+1}$, the x-axis and the ordinates $x = 0$ and $x = 1$.

Solution

$$\text{Required area} = \int_0^1 y \cdot dx$$

$$= \int_0^1 \frac{x^2}{x^2+1} dx = \int_0^1 \frac{x^2+1-1}{x^2+1} dx$$

$$= \int_0^1 1 \cdot dx - \int_0^1 \frac{1}{x^2+1} dx$$

$$= (x - \tan^{-1} x)\Big]_0^1$$

$$= (1 - \tan^{-1} 1) - (0 - \tan^{-1} 0)$$

$$= 1 - \frac{\pi}{4} \text{ sq. units.}$$

Example 3 Find the area enclosed by the curve $y = 4x - x^2 - 3$ and the x-axis.

Solution

We shall first find the points of intersection of the curve and the x-axis.

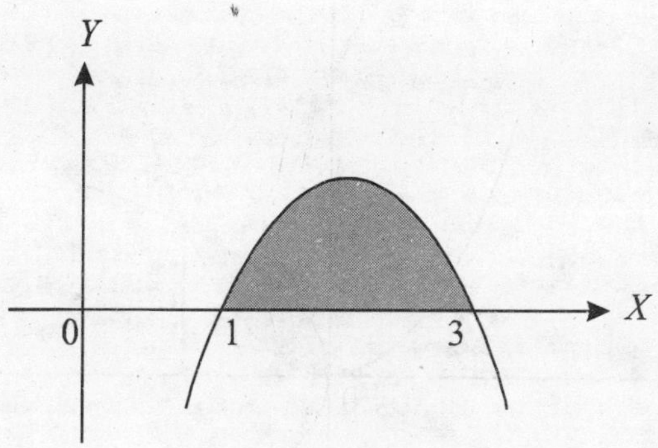

Fig. 2.4

On the x-axis, $y = 0$ substituting $y = 0$ in the given equation, we get $0 = 4x - x^2 - 3$

i.e., $\quad x^2 - 4x + 3 = 0$

i.e., $\quad (x - 1)(x - 3) = 0$

$\therefore \quad x = 1$ or $x = 3$

$\therefore$ Area bounded by the curve and the x-axis is

$$= \int_1^3 y \cdot dx = \int_1^3 (4x - x^2 - 3)dx$$

$$= \left[4\frac{x^2}{2} - \frac{x^3}{3} - 3x\right]_1^3 = \left[2x^2 - \frac{x^3}{3} - 3x\right]_1^3$$

$$= \left[2 \cdot 3^2 - \frac{3^3}{3} - 3 \cdot 3\right] - \left[2 \cdot 1 - \frac{1^3}{3} - 3 \cdot 1\right]$$

$$= [18 - 9 - 9] - \left[2 - \frac{1}{3} - 3\right]$$

$$= -\left[-1 - \frac{1}{3}\right] = \frac{4}{3} \text{ sq. units.}$$

Example 4 Find the area enclosed by the curve $y^2 = 4x$ and $x^2 = 4y$
We shall first find the points of intersection of the curves.

Solution

Substituting $y = \dfrac{x^2}{4}$ in $y^2 = 4x$ we get $\left[\dfrac{x^2}{4}\right]^2 = 4x$ or $\dfrac{x^4}{16} = 4x$

$$x^4 - 64x = 0$$

$$\therefore \quad x(x^3 - 64) = 0$$

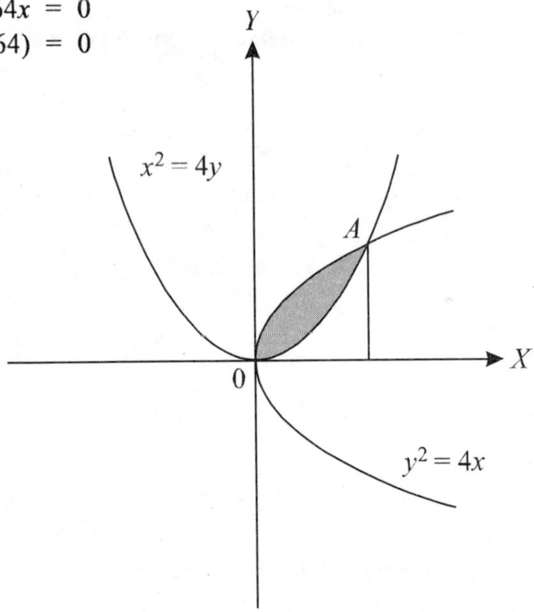

Fig. 2.5

$\therefore \quad x = 0 \quad \text{or} \quad x^3 - 64 = 0$

$\therefore \quad x = 0 \quad \text{or} \quad x = 4$

Area enclosed between the curves

$$= \int_0^4 \left[\sqrt{4x} - \frac{x^2}{4}\right] dx = \left[2 \cdot \frac{x^{3/2}}{3/2} - \frac{x^3}{12}\right]_0^4$$

$$= \left[\frac{4}{3} \cdot 4^{3/2} - \frac{4^3}{12}\right] - 0 = \frac{32}{3} - \frac{16}{3} = \frac{16}{3} \text{ sq. units}$$

Example 5 Find the area enclosed between the curve $y^2 = 4x$ and the line $x = 2y$.

Solution

Solving the given equation for x, we get

$$\left[\frac{x}{2}\right]^2 = 4x \quad \therefore \frac{x^2}{4} = 4x$$

$\therefore \quad x^2 = 16x$

$x^2 - 16x = 0$

i.e., $\quad x(x - 16) = 0$

$\therefore \quad x = 0 \quad \text{or} \quad x = 16$

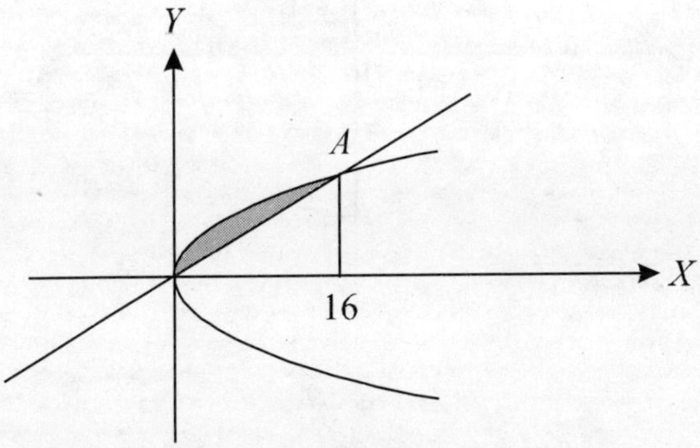

Fig. 2.6

Area enclosed between the curve and the line

$$= \int_0^{16} \left[\sqrt{4x} - \frac{x}{2}\right] dx$$

$$= \left[2 \cdot \frac{x^{3/2}}{3/2} - \frac{x^2}{4}\right]_0^{16}$$

$$= \left[\frac{4}{3}\cdot 16^{3/2} - \frac{16^2}{4}\right] - 0$$

$$= \frac{256}{3} - 64 = \frac{64}{3} \text{ sq. units}$$

Example 6 Find the area of the circle $x^2 + y^2 = a^2$ by integration.
Area of the circle is equal to four times the area of the circle in the first quadrant. Since the circle is symmetric about both x and y axis.

Solution

Area of circle = 4 Area OAB

Now $y^2 = a^2 - x^2$ or $y = \sqrt{a^2 - x^2}$

Clearly $O = (0,0)$ and $A = (a,0)$

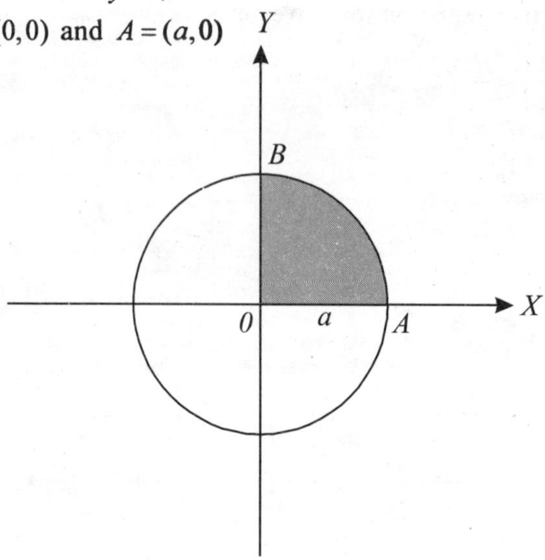

Fig. 2.7

$$\therefore \quad \text{Area } OAB = \int_0^a y \cdot dx = \int_0^a \sqrt{a^2 - x^2} \cdot dx$$

Substituting $x = a\sin\theta$, so that $dx = \cos\theta d\theta$

Now $\theta = \sin^{-1}\frac{x}{a}$

when $x = 0,\ \theta = \sin^{-1}\frac{0}{a} = \sin^{-1} 0 = 0$

when $x = a,\ \theta = \sin^{-1}\frac{a}{a} = \sin^{-1} 1 = \frac{\pi}{2}$

$$\therefore \quad \text{Area } OAB = \int_0^{\frac{\pi}{2}} \sqrt{a^2 - a^2 \sin^2\theta} \cdot a\cos\theta \cdot d\theta$$

$$= \int_0^{\frac{\pi}{2}} a\cos\theta \cdot a\cos\theta \cdot d\theta = a^2 \int_0^{\frac{\pi}{2}} \cos^2\theta \cdot d\theta$$

$$= a^2 \int_0^{\frac{\pi}{2}} \frac{1+\cos 2\theta}{2} d\theta = \frac{a^2}{2}\left[\left(\theta + \frac{\sin 2\theta}{2}\right)\right]_0^{\frac{\pi}{2}}$$

$$= \frac{a^2}{2}\left[\left(\frac{\pi}{2} + \frac{\sin\pi}{2}\right) - \left(0 + \frac{\sin 0}{2}\right)\right] = \frac{\pi a^2}{4}$$

$$\therefore \quad \text{Area of the circle} = 4 \cdot \frac{\pi a^2}{4} = \pi a^2 \quad \text{sq. units.}$$

Example 7 Prove that the area of the ellipse $\frac{x^2}{a^2} + \frac{y^2}{b^2} = 1$ is πab sq. units.

Solution

Since the ellipse is symmetric about the coordinate axes, area of the ellipse is equal to 4 times the area of the ellipse in the first quadrant.

Since $O = (0, 0)$, A = $(a, 0)$ the limits are from 0 to a.

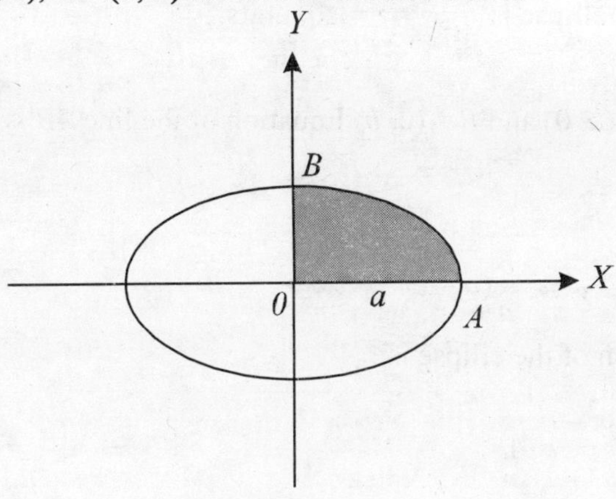

Fig. 2.8

Solution Also $\quad \frac{y^2}{b^2} = 1 - \frac{x^2}{a^2}$

i.e., $$y^2 = \frac{b^2}{a^2}(a^2 - x)$$

$\therefore$ $$y = \frac{b}{a}\cdot\sqrt{a^2 - x^2}$$

$\therefore$ $$\text{Area } OB = \int_0^a y\cdot dx = \int_0^a \frac{b}{a}\cdot\sqrt{a^2 - x^2}\cdot dx = \frac{b}{a}\int_0^a \sqrt{a^2 - x^2}\cdot dx$$

$$= \frac{b}{a}\cdot\frac{\pi a^2}{4}$$ as proceeding in the above example.

$$= \frac{\pi ab}{4}$$

$\therefore$ Area of the ellips $= 4\cdot\frac{\pi ab}{4} = \pi ab$ sq. units.

Example 8 AOB is the positive quadrant of the ellipse $\frac{x^2}{a^2} + \frac{y^2}{b^2} = 1$ in which $OA = a$, $OB = b$. Show that the area between the chord AB and the arc AB of the ellipse is $\frac{(\pi - 2)ab}{4}$ sq. units.

Solution

$\Rightarrow$ Now $A = (a, 0)$ and $B = (0, b)$ Equation of the line AB is

$$\frac{x}{a} + \frac{y}{b} = 1$$

$\therefore$ $$y = \frac{b}{a}(a - x) \qquad \text{... i}$$

Also, equation of the ellipse is

$$\frac{x^2}{a^2} + \frac{y^2}{b^2} = 1$$

$\therefore$ $$y = \frac{b}{a}\sqrt{a^2 - x^2} \qquad \text{.... ii}$$

$$\text{Let } A_1 = \text{Area } OAPB = \int_0^a \frac{b}{a}\sqrt{a^2 - x^2}\,dx$$

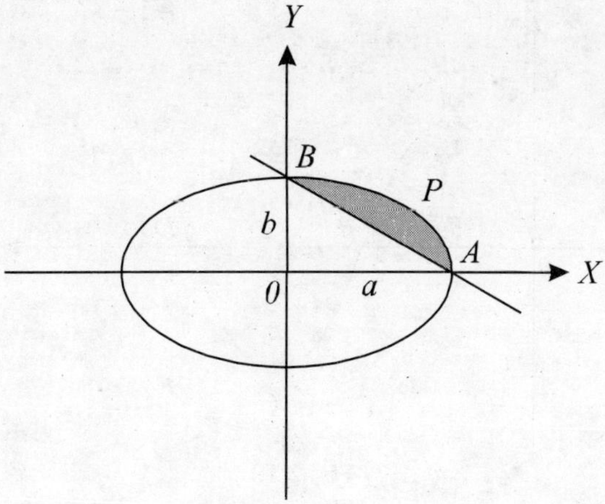

Fig. 2.9

$$= \frac{b}{a}\int_0^a \frac{x}{2}\sqrt{a^2 - x^2} + \frac{a^2}{2}\sin^{-1}\frac{x}{a}\Bigg]_0^a$$

$$= \frac{b}{a}\left[\left(0 + \frac{a^2}{2}\cdot\sin^{-1}1\right) - (0)\right]$$

$$= \frac{b}{a}\cdot\frac{a^2}{2}\cdot\frac{\pi}{2} = \frac{\pi ab}{4}$$

and $$A_2 = \text{Area } OAB = \int_0^a \frac{b}{a}(a - x)\,dx$$

$$= \frac{b}{a}\left(ax - \frac{x^2}{2}\right)\Bigg]_0^a = \frac{b}{a}\left(a^2 - \frac{a^2}{2}\right) = \frac{b}{a}\cdot\frac{a^2}{2} = \frac{ab}{2}$$

Required area $= A_1 - A_2$

$$= \frac{\pi ab}{4} - \frac{ab}{2} = \frac{(\pi - 2)ab}{4} \text{ sq. units}$$

Example 9 Find the area between the curve $y = x^3 - 6x^2 + 8x$ and the x-axis.

Solution:

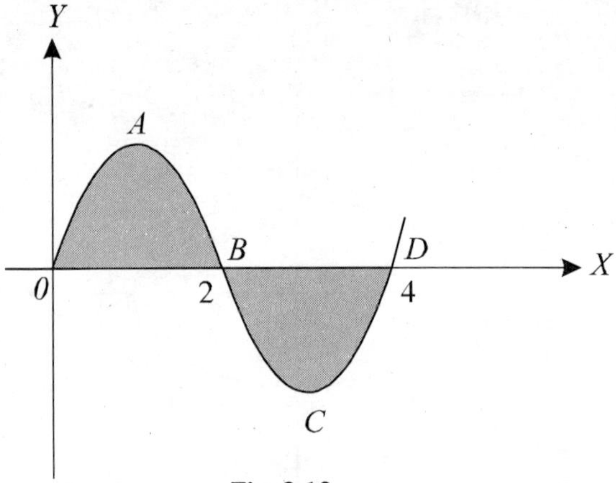

Fig. 2.10

On the x – axis, $y = 0$. Substituting $y = 0$ the given equation reduces to

$$x^3 - 6x^2 + 8x = 0$$

i.e., $$x(x^2 - 6x + 8) = 0$$

i.e., $$x(x-2)(x-4) = 0$$

$\therefore$ $$x = 0 \text{ or } x = 2 \text{ or } x = 4.$$

The curve intersects the x – axis at $x = 0$, $x = 2$ and $x = 4$. Rough sketch of the curve is shown in figure 4.10

$$\text{Area } OAB = \int_0^2 \left(x^3 - 6x^2 + 8x\right) dx = \left.\frac{x^4}{4} - 6\cdot\frac{x^3}{3} + 8\cdot\frac{x^2}{2}\right]_0^2 = 4$$

$$\text{Area } OAB = \int_0^2 \left(x^3 - 6x^2 + 8x\right) dx = \left.\frac{x^4}{4} - 6\cdot\frac{x^3}{3} + 8\cdot\frac{x^2}{2}\right]_0^2 = -4$$

$$\text{Required area} = 4 + |-4| = 8 \text{ sq. units.}$$

EXERCISES

1. **Find the area bounded by the curve, x - axis and the ordiantes given**

   a. $y = x^2 - 3x + 7; x = 1, x = 3$ $\left[\textbf{Ans: } \frac{32}{3} \text{sq.units}\right]$

   b. $y = 10x^2; x = 1, x = 2$ $\left[\textbf{Ans: } \frac{70}{3} \text{sq.units}\right]$

c. $y = x^2 - 3x - 3; x = -1, x = 2$ $\left[\textbf{Ans}: \frac{21}{3} \text{sq.units}\right]$

d. $y = x^2 + 1; x = 0, x = 1$ $\left[\textbf{Ans}: \frac{4}{3} \text{sq.units}\right]$

e. $y = 4x^2; x = 0, x = 1$ $\left[\textbf{Ans}: \frac{4}{3} \text{sq.units}\right]$

f. $y = x^3; x = 1, x = 2$ $\left[\textbf{Ans}: \frac{15}{3} \text{sq.units}\right]$

2. Find the area bounded by the Parabola $x^2 = 4ay$, x –axis and $x = 0, x = 4a$

[**Ans**: $4 \log 2$ sq.units]

4. Find the area bounded by each of the following curves and *x*-axis

a. $y = 4x - x^2$ $\left[\textbf{Ans}: \frac{32}{3} \text{sq.units}\right]$

b. $y = x^2 - 7x + 10$ $\left[\textbf{Ans}: \frac{9}{2} \text{sq.units}\right]$

c. $y = x^2 - 5x + 6$ $\left[\textbf{Ans}: \frac{1}{6} \text{sq.units}\right]$

d. $y = x^2 - 3x + 2$ $\left[\textbf{Ans}: \frac{1}{6} \text{sq.units}\right]$

e. $y = x^2 - 3x$ $\left[\textbf{Ans}: \frac{9}{2} \text{sq.units}\right]$

5. Find the area enclosed between

a. $y^2 = x \; ; \; x^2 = y$ $\left[\textbf{Ans}: \frac{1}{3} \text{sq.units}\right]$

b. $y^2 = ax \; ; \; x^2 = ay$ $\left[\textbf{Ans}: \frac{a^2}{3} \text{sq.units}\right]$

c. $y^2 = 6x \; ; \; y = x$ [**Ans**: 12sq.units]

d. $y^2 = 4x \; ; \; x = 2y$ $\left[\textbf{Ans}: \frac{64}{3} \text{sq.units}\right]$

e. $y^2 = 4x$; $y = x$ $\left[\textbf{Ans}: \frac{8}{3}\text{sq.units}\right]$

f. $y = 8 - x^2$; $y = x^2$ $\left[\textbf{Ans}: \frac{64}{3}\text{sq.units}\right]$

g. $y^2 = x$; $x = y$ $\left[\textbf{Ans}: \frac{1}{6}\text{sq.units}\right]$

6. **Find the area of the circle $x^2 + y^2 = 9$ by integration**

$[\textbf{Ans}: 9\pi \text{ sq.units}]$

7. **Find the area of ellipse $\frac{x^2}{9} + \frac{y^2}{4} = 1$ by integration**

$[\textbf{Ans}: 6\pi \text{ sq.units}]$

2.6 Volumes of Solids of Revolution

2.6.1 Solid of revolution

Let *AB* be an arc of a curve and *CD* be a straight line which does not intersect the arc *AB*. Let *AA*' and *BB'* be perpendiculars to the line *CD*. If the region *AA 'B' BA* is rotated or revolved about the line *CD* a solid will be obrtained which is called a solid of revolution. This solid is said to be obtained by the revolution of the arc *AB* about the line *CD*. The straight line *CD* is called the axis of revolution.

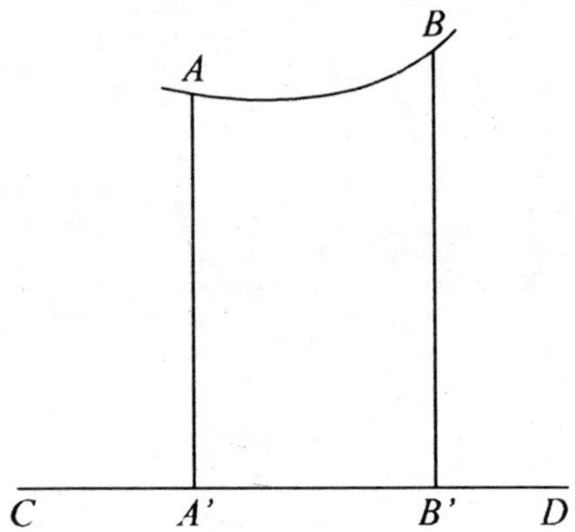

Fig. 2.11

Definition

A solid of revolution isgeneraed by revolving a plane area *R* about a line *L* in the plane. Line *L* is known as the axis of revolution. Line *L* does not intersect the plane area *R* but may touch the boundary of *R*.

Examples

i. If a rectangle is revolved about one of its sides, we get a right circular cylinder as the solid or revolution.

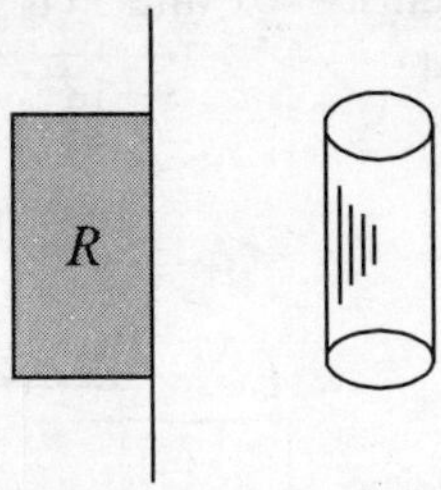

ii. If a right - angled triangle is revolved about one of its sides containing the right angle, we get a right circular cone.

iii. If a semi - circle is revolved about its diameter, we get a sphere.

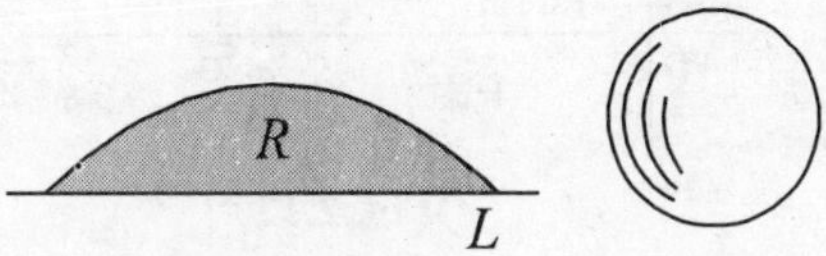

iv. If a parabola is revolved about x - axis, we get a paraboloid. If the axis of revolution crosses the arc to be revolved about it, the two parts of the arc will revolve through some common space and the solid of revolution will have an overlapping volume if the two parts are not symmetrical about the axis of revolution. When the arc is symmetrical about the axis, each part generates the same solid of revolution.

2.6.2 Volume of a solid of revolution about the x - axis

Theorem : Volume of the solid generated by the revolution of the curve $y = f(x)$ between $x = a$ and $x = b$ about the x – axis is.

$$\int_a^b \pi y^2 dx$$

(It begin assumed that the curve does not intersect the x - axis)

Proof : Let $y = f(x)$ be the equation of the curve AB.

Let AL and BM be the ordinates $x = a$ and $x = b$ respectively.

Let $P(x, y)$ be a variable point on the curve. Let V changes with it. So, V is function of x. Let $Q(x + \delta x, y)$ be another point on the curve near the point P, such that y constantly increases (or decreases) between P and Q. Complete the rectangles $PSTQ'$ and $P'STQ$. Denote by $V + \delta V$, the volume of the solid generated by the area $ALTQ$. The δV is the volume generated by the rectangular area $PSTQ$. Obviously δV lies between the volumes of the right-circular cylinders generated by the rectangles $PSTQ'$ and $P'STQ$.

i.e., δV lies between $\pi y^2\, \delta x$ and $\pi(y + \delta y)^2 \delta x$ [$\because$ volume of the right circular cylinder is $\pi r^2 h$]

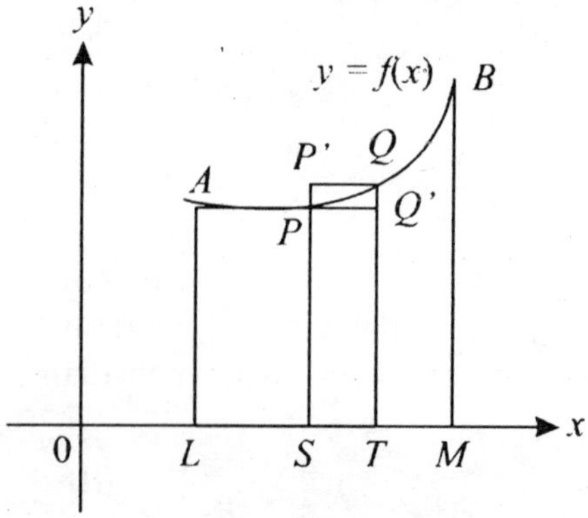

Fig. 2.12

or $\dfrac{\delta V}{\delta x}$ lies between πy^2 and $\pi(y + \delta y)^2$. Proceeding to the limit, when $\delta x \to 0$ and consequently $\delta y \to 0$, So $\pi(y + \delta y)^2 \to \pi y^2$

$$\underset{\delta x \to 0}{lt} \frac{\delta V}{\delta x} = \pi y^2. \quad \text{But} \quad \underset{\delta x \to 0}{lt} \frac{\delta V}{\delta x} = \frac{dV}{dx} \quad \therefore \quad \frac{dV}{dx} = \pi y^2$$

$$\int_a^b \pi y^2 dx = \int_a^b \frac{dV}{dx} dx = (V)_a^b$$

= (Value of V at $x = b$) – (Value of V at $x = a$)
= (Value of V at $x = b$) – 0 (as the volume for $x = a$ is to be taken as 0)
= Volume of the solid generated by the revolution of the area $ALMB$

$$\therefore \text{ Required volume} = \int_a^b \pi y^2 dx$$

Note : Similarly, the volume of the solid generated by the revolution of the area bounded by the curve $x = f(y)$, the y – axis and the lines $y = a, y = b$ about the y – axis is

$$\int_a^b \pi y^2 dy$$

WORKED EXAMPLES

Example 1 Find the volume of the solid, obtained by rotating the portion of the parabola $y = x^2$ between $x = 0$ and $x = 2$ about the x – axis.

Solution

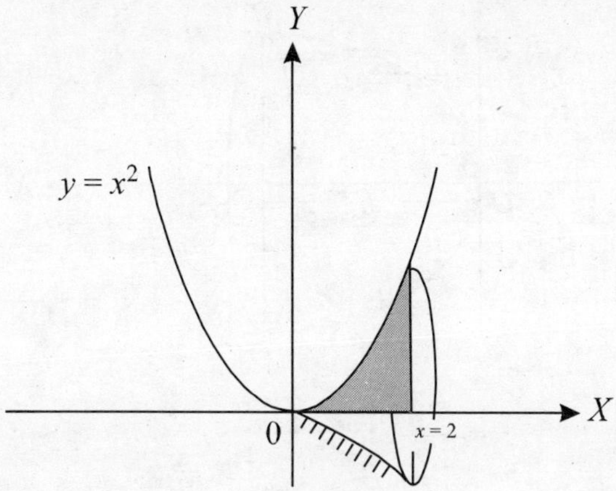

Fig. 2.13

Equation of the parabola is $y = x^2$

The volume generated when the parabola revolves about the x – axis.

$$= \pi\int_0^2 y^2 dx = \pi\int_0^2 x^4 dx = \pi\left(\frac{x^5}{5}\right)_0^2 = \frac{32\pi}{5} \text{ cubic units}$$

Example 2 Find the volume of the solid, formed by the revolution of the area bounded by the parabola $y^2 = 25x$, the x – axis and the lines $x = 1$, $x = 2$ about the x – axis.

Solution

Equation of the parabola is $y^2 = 25x$

$$\therefore \quad \text{Required volume} = \int_1^2 \pi y^2 dx = \pi\int_1^2 25x dx = 25\pi\left(\frac{x^2}{2}\right)_1^2$$

$$= \frac{25}{2}\pi(4-1) = \frac{75}{2} \text{ cubic units.}$$

Example 3 The area bounded by $y^2 = 4ax$, the x – axis and the ordinates at $x = 0$ and $x = h$ rotates about the x – axis. Show that the volume generated is $2\pi ah^2$.

(or) Find the volume of the paraboloid generated by revolving the parabola $y^2 = 4ax$ about x – axis from $x = 0$ to $x = h$.

Solution

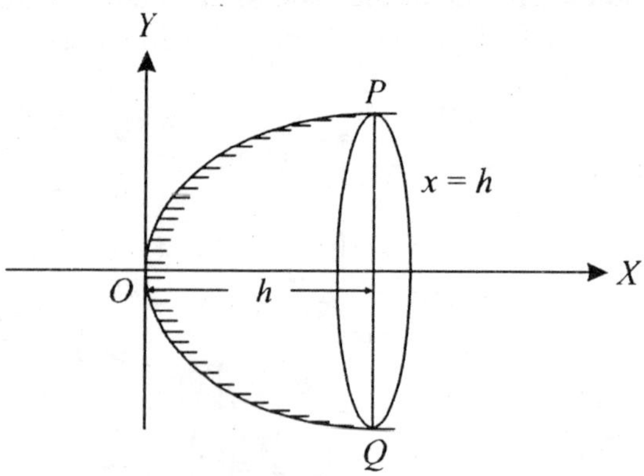

Fig. 2.14

The given curve is $y^2 = 4ax$

$$\therefore \quad \text{Required volume} = \int_0^h \pi y^2 dx = \int_0^h \pi(4ax)dx$$

$$= 4a\pi \int_0^h x dx = 4a\pi \left(\frac{x^2}{2}\right)_0^h$$

$$= 4a\pi\left(\frac{h^2}{2} - 0\right) = 2\pi a h^2 \text{ cubic units.}$$

Example 4 Find the volume generated when the area bounded by $y^2 = x^3$ and $x = 2$ revolves about the x – axis.

Solution

The curve $y^2 = x^3$ is symmetrical about x – axis, since the power of y is even. Hence, it is enough if we revolve the area about the x – axis. One limit is given to be $x = 2$. The curve passes through the origin. Therefore, we can take the limit to be $x = 0$.

$$\text{Thus the required volume} = \int_0^2 \pi y^2 dx = \pi \int_0^2 x^3 dx = \pi\left(\frac{x^4}{4}\right)_0^2$$

$$= \frac{\pi}{4}(16 - 0) = 4\pi \text{ cubic units.}$$

Example 5 The area bounded by $y=(x+2)(1-x)$ is revolved about the x – axis. Find the volume generated.

Solution The intersections of the curve with x – axis are given by solving it with $y=0$ i.e., by $0=(x+2)(1-x)$

Hence $x=-2$ or 1. These are the limits of integration.

$$\therefore \quad \text{Required volume} = \int_{-2}^{1} \pi y^2 dx = \pi \int_{-2}^{1} [(x+2)(x-1)]^2 dx$$

$$= \pi \int_{-2}^{1} (2-x-x^2)^2 dx = \pi \int_{-2}^{1} (4-4x-3x^2+2x^3+x^4) dx$$

$$= \pi \left[4x - 4\frac{x^2}{2} - 3\cdot\frac{x^3}{3} + 2\cdot\frac{x^4}{4} + \frac{x^5}{5} \right]_{-2}^{1}$$

$$= \pi \left[4x - 2x^2 - x^3 + \frac{x^4}{2} + \frac{x^5}{5} \right]_{-2}^{1}$$

$$= \pi \left[\left(4-2-1+\frac{1}{2}+\frac{1}{5} \right) - \left(-8-8+8+8-\frac{32}{5} \right) \right]$$

$$= \pi \left[\frac{17}{10} + \frac{32}{5} \right] = \frac{81\pi}{10} \text{ cubic units.}$$

Example 6 Find the volume of the solid obtained by rotating the area bounded by the curve $\sqrt{x}+\sqrt{y}=1$ and in the first quadrant about the y – axis.

Solution

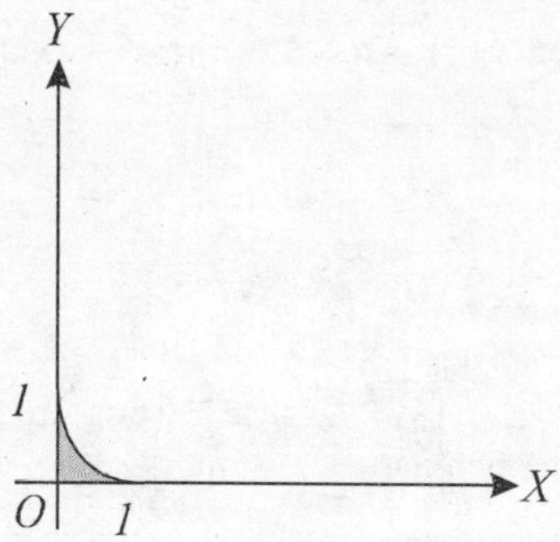

Fig. 2.15

Equation of the curve is $\sqrt{x}+\sqrt{y}=1$ $\therefore \sqrt{x}=1-\sqrt{y}$ or $x=\left(1-\sqrt{y}\right)^2$

To find where $\sqrt{x}=1-\sqrt{y}$ cuts the $y-$ axis,

we have to solve $\sqrt{x}=1-\sqrt{y}$ and $x=0$

$\therefore\ 1-\sqrt{y}=0 \Rightarrow y=1$

Similarly the curve cuts the $x-$ axis $x=1$

$$\therefore \text{ Required volume } = \int_0^1 \pi x^2 dy = \pi\int_0^1\left(1-\sqrt{y}\right)^4 dy$$

$$= \pi\int_0^1\left(1+y-2\sqrt{y}\right)^2 dy$$

$$= \pi\int_0^1\left(1+6y+y^2-4y^{\frac{1}{2}}-4y^{\frac{3}{2}}\right)dy$$

$$= \pi\left[y+6\frac{y^2}{2}+\frac{y^3}{2}-4\cdot\frac{2}{3}y^{\frac{3}{2}}-\frac{2}{5}4y^{\frac{5}{2}}\right]_0^1$$

$$= \pi\left[y+3y^2+\frac{y^3}{2}-\frac{8}{3}y^{\frac{3}{2}}-\frac{8}{5}y^{\frac{5}{2}}\right]_0^1$$

$$= \pi\left[1+3+\frac{1}{3}-\frac{8}{3}-\frac{8}{5}\right] = \pi\left[\frac{13}{3}-\frac{8}{3}-\frac{8}{5}\right]$$

$$= \pi\left[\frac{5}{3}-\frac{8}{3}\right] = \pi\left[\frac{25-24}{15}\right] = \frac{\pi}{15} \text{ cubic units.}$$

Example 7 Find the volume of the solid generated by the rotation of the area bounded by the curve $y=a\cos x$, the x – axis and the lines $x=0$, $x=\frac{\pi}{2}$ about the $x-$ axis.

Solution

Give curve is $y=a\cos x$

$$\therefore \text{ Required volume } = \int_0^{\frac{\pi}{2}} \pi y^2 dx = \pi\int_0^{\frac{\pi}{2}} a^2\cos^2 x dx$$

$$= \frac{\pi a^2}{2}\int_0^{\frac{\pi}{2}}(1+\cos 2x)dx$$

$$= \frac{\pi a^2}{2}\left[x + \frac{\sin 2x}{2}\right]_0^{\frac{\pi}{2}} = \frac{\pi a^2}{2}\left[\frac{\pi}{2} + \frac{\sin \pi}{2} - 0\right]$$

$$= \frac{\pi a^2}{2}\left[\frac{\pi}{2} + 0\right] = \frac{\pi^2 a^2}{4} \text{ cubic units.}$$

Example 8 Obtain the volume of sphere of radius r using integration.

Solution

Let the sphere generated by the revolution of the semi-circle ABC, of radius r about its diameter CA.

Taking CA as the X – axis and its mind point O as the origin, the equation of the circle ABC is

$$x^2 + y^2 = r^2 \Rightarrow y^2 = r^2 - x^2$$

Hence the required volume = $\int_{-r}^{r} \pi y^2 dx = \pi \int_{-r}^{r} (r^2 - x^2) dx$

$$= 2\pi \int_0^r (r^2 - x^2) dx \quad [\because (r^2 - x^2) \text{ is even function}]$$

$$= 2\pi\left(r^2 x - \frac{x^3}{3}\right)_0^r = 2\pi\left(r^3 - \frac{r^3}{3}\right)$$

$$= 2\pi \cdot \frac{2r^3}{3} = \frac{4}{3}\pi r^3$$

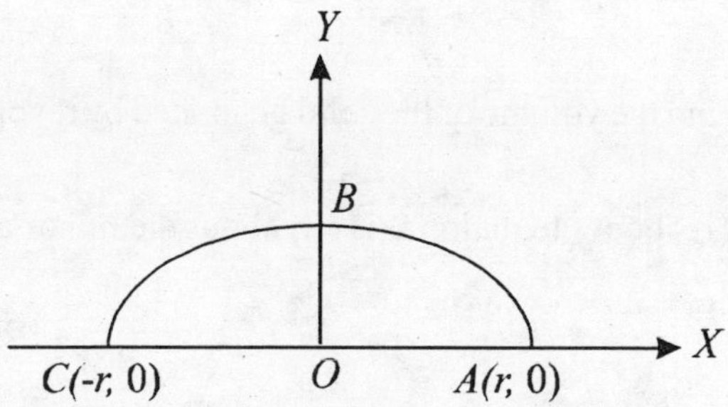

Fig. 2.16

Example 9 Find the volume of a cone of height h and base radius r using integration.

Solution

Consider the right angled triangle OAB with $OA = h$ and $AB = r$. The right circular cone of height h and radius r is generated by the revolution of the area OAB about the x – axis.

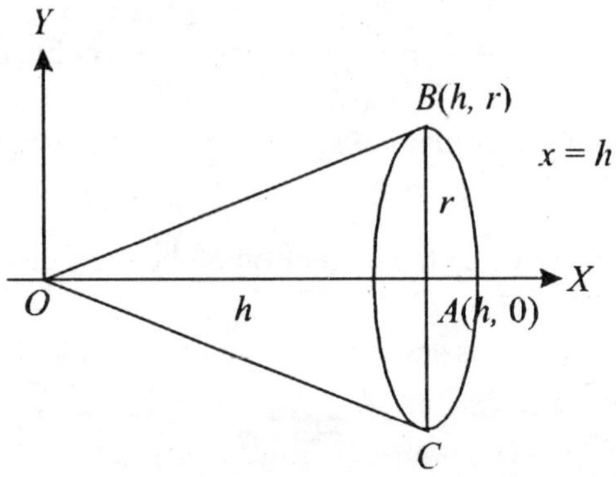

Fig. 2.17

Equation of the hypotenuse (i.e., OB) is $\dfrac{y-0}{r-0} = \dfrac{x-0}{h-0}$ i.e., $\dfrac{y}{r} = \dfrac{x}{h}$

i.e., $$y = \left(\frac{r}{h}\right)x$$

$\therefore$ Volume of the cone $$= \int_0^h \pi y^2 dx = \pi \int_0^h \frac{r^2}{h^2} x^2 dx = \pi \frac{r^2}{h^2}\left(\frac{x^3}{3}\right)_0^h$$

$$= \pi \frac{r^2}{h^2} \cdot \frac{h^3}{3} = \frac{1}{3}\pi r^2 h$$

Example 10 Find the volume of the solid generated by revolving the ellipse $\dfrac{x^2}{a^2} + \dfrac{y^2}{b^2} = 1$ (i) about the major axis (ii) about the minor axis

Solution

i. Equation of the ellipse is $\dfrac{x^2}{a^2} + \dfrac{y^2}{b^2} = 1 \Rightarrow \dfrac{y^2}{b^2} = 1 - \dfrac{x^2}{a^2} = \dfrac{a^2 - x^2}{a^2}$

$\therefore$ $$y^2 = \frac{b^2}{a^2}(a^2 - x^2)$$

$$\text{Required volume} = \int_{-a}^{a} \pi y^2 dx = \pi \int_{-a}^{a} y^2 dx$$

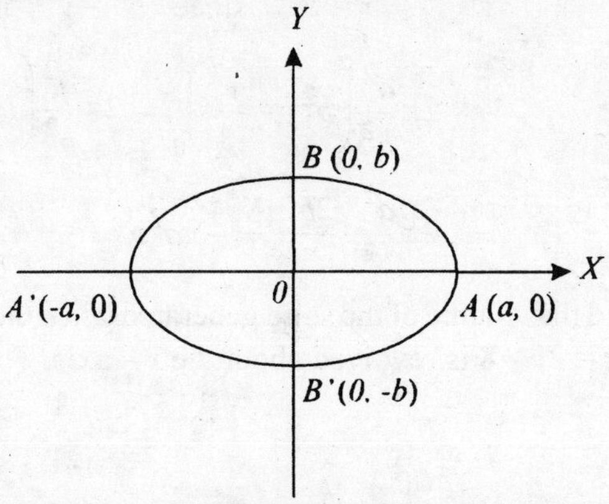

Fig. 2.18

$$= \pi \int_{-a}^{a} \frac{b^2}{a^2}(a^2 - x^2)dx$$

$$= \pi \frac{b^2}{a^2} \int_{-a}^{a} (a^2 - x^2)dx = 2\pi \frac{b^2}{a^2} \int_{0}^{a} (a^2 - x^2)dx$$

[∵ integrand is even function]

$$= 2\pi \frac{b^2}{a^2}\left(a^2 x - \frac{x^3}{3}\right)_0^a = 2\pi \frac{b^2}{a^2}\left(a^3 - \frac{a^3}{3}\right)$$

$$= 2\pi \frac{b^2}{a^2} \cdot \frac{2a^3}{3} = \frac{4}{3}\pi ab^2$$

ii. Equation of the ellipse is $\frac{x^2}{a^2} + \frac{y^2}{b^2} = 1 \Rightarrow \frac{x^2}{a^2} = 1 - \frac{y^2}{b^2} = \frac{b^2 - y^2}{b^2}$

$$\therefore \quad x^2 = \frac{a^2}{b^2}(b^2 - y^2)$$

$$\therefore \quad \text{Required volume} = \int_{-b}^{b} \pi x^2 dy = \pi \int_{-b^2}^{b} \frac{a^2}{b^2}(b^2 - y^2)dy$$

$$= \pi\frac{a^2}{b^2}2\int_0^b (b^2 - y^2)dy,$$

since $(b^2 - y^2)$ is even function

$$= 2\pi\frac{a^2}{b^2}\left(b^2 y - \frac{y^3}{3}\right)_0^b = 2\pi\frac{a^2}{b^2}\left(b^3 - \frac{b^3}{3}\right)$$

$$= 2\pi\frac{a^2}{b^2}\cdot\frac{2b^3}{3} = \frac{4}{3}\pi a^2 b$$

Example 11 Find the volume of the solid generated when the region bouned by $x^2 = 8y$ and $x = 2y - 8$ is revolved about the x – axis.

Solution

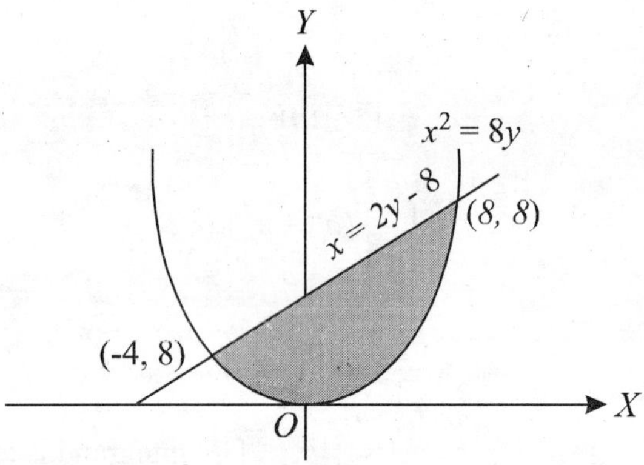

Fig. 2.19

Equation of the parabola is $x^2 = 8y$ ... 1

Equation of thestraight line is $x = 2y - 8$ ... 2

For finding the points of intersection, solve (1) and (2)

Putting the valueof $x = 2y - 8$ in (1), we get

$(2y-8)^2 = 8y$ or $4y^2 + 64 - 32y = 8y$

or $(y-8)(y-2) = 0$ $\therefore$ $y = 2, y = 8$

From (2), when $y = 2, x = 4 - 8 = -4$

when $y = 8, x = 16 - 8 = 8$

Thus, the curves intersect at (–4, 2) and (8, 8) as shown in figure

$\therefore$ Required volume = volume generated by shaded portion

$$= \int_{-4}^{8} \pi\ [(y \text{ of upper curve})^2 - (y \text{ of lower curve})^2]dx$$

$$= \pi \int_{-4}^{8} \left(\frac{(x+8)^2}{4} - \frac{x^4}{64} \right) dx$$

$$= \pi \int_{-4}^{8} \left[\frac{1}{4}(x^2 + 64 + 16x) - \frac{1}{64}x^4 \right] dx$$

$$= \pi \left[\frac{1}{4}\left(\frac{x^3}{3} + 64x + 16\frac{x^2}{2} \right) - \frac{1}{64}\frac{x^5}{5} \right]_{-4}^{8}$$

$$= \pi \left[\frac{1}{4}\left(\frac{8^3}{3} + 8^3 + 8^3 \right) - \frac{1}{64}\frac{8^5}{5} - \frac{1}{4}\left(-\frac{4^3}{3} - 4^4 + \frac{4^4}{2} \right) - \frac{1}{64} \times \frac{4^5}{5} \right]$$

$$= \pi \left[\frac{8^3}{4}\left(\frac{1}{3} + 2 \right) - \frac{8^3}{5} - \frac{4^3}{4}\left(\frac{-1}{3} - 4 + \frac{4}{2} - \frac{4^2}{5} \right) \right]$$

$$= \pi \left[\frac{8^3}{4}\left(\frac{32}{15} \right) - 4^2\left(\frac{-7}{3} \right) - \frac{4^2}{5} \right]$$

$$= \pi \left[\frac{8^4}{15} + 4^2 \cdot \frac{33}{15} \right]$$

$$= \frac{\pi}{15}(4096 + 528)$$

$$= \frac{4624\pi}{15} = 308.2\,\pi \text{ cubic units.}$$

Example 12 Find the volume generaed by revolving about y – axis the area bounded by the curve $y = x^3$, the line $y - 1$, and between $x = 0$ and $x = 1$.

Solution

Given curve is $y = x^3$, $\therefore\ x = y^{\frac{1}{3}}$

Required volume = Volume generated by shaded portion

$$= \int_0^1 \pi x^2 dy = \pi \int_0^1 y^{\frac{2}{3}} dy = \pi \cdot \frac{3}{5}\left[y^{\frac{5}{3}} \right]_0^1$$

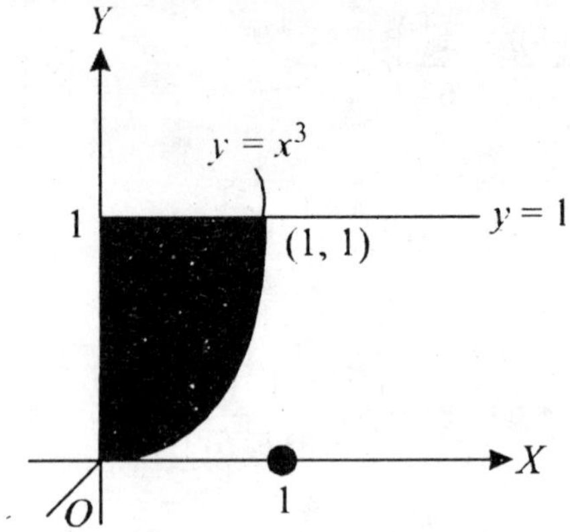

Fig. 2.20

$$= \frac{3\pi}{5}(1-0) = \frac{3\pi}{5} \text{ cubic units.}$$

Note: About x – axis

$$\text{Required volume} = \int_0^1 \pi\,[(y \text{ of upper curve})^2 - (y \text{ of lower curve})^2]dx$$

$$= \pi\int_0^1 (1^2 - x^6)dx = \pi\left(x - \frac{x^7}{7}\right)_0^1$$

$$= \pi\left(1 - \frac{1}{7}\right) = \frac{6\pi}{7} \text{ cubic units.}$$

Example 13 The area between the parabola $y^2 = 9x$ and $x^2 = 9y$ revolves about the x – axis. Find the volume of the solid so generated.

Solution

Given equations of the parabolas are

$$y^2 = 9x \qquad \ldots 1$$

$$x^2 = 9y \qquad \ldots 2$$

Solve (1) and (2) for the points of intersection of two curves.

Putting the value of $y = \dfrac{x^2}{9}$ from (2) in (1), we get

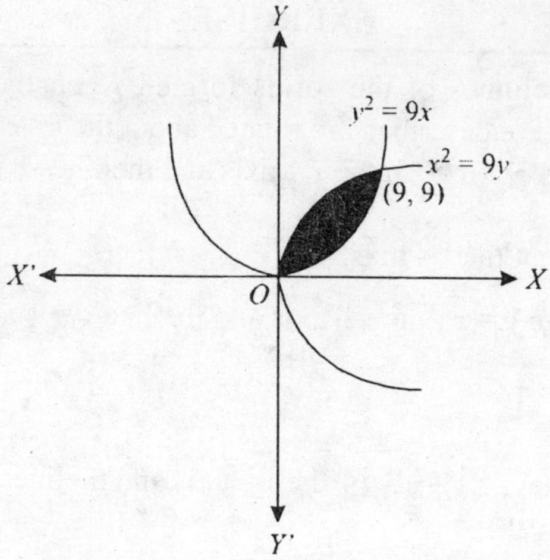

Fig. 2.21

$$\frac{x^4}{9^2} = 9x \Rightarrow x^4 - 9^3 x = 0 \text{ or } x(x^3 - 9^3) = 0 \Rightarrow x = 0,\ x = 9$$

When $x = 0$, $y = 0$, and when $x = 9$, $y = 9$

Thus, the curves intersect at (0, 0) and (9, 9) as shown in figure.

$\therefore$ Required volume = Volume generated by shaded portion

$$= \int_0^9 \pi\, [(y \text{ of upper curve})^2 - (y \text{ of lower curve})^2]dx$$

$$= \pi \int_0^9 \left(9x - \frac{x^4}{81}\right) dx$$

$$= \pi \left(9\frac{x^2}{2} - \frac{x^5}{5 \times 81}\right)_0^9$$

$$= \pi \left(\frac{9^3}{2} - \frac{9^5}{5 \times 81}\right)$$

$$= \pi \cdot 9^3 \left(\frac{1}{2} - \frac{1}{5}\right)$$

$$= 729\pi \left(\frac{3}{10}\right) = 218.7\ \pi \text{ cubic units}$$

EXERCISES

1. Find the volumes of the solids formed when the regions whose boundaries are described are rotated about the given axis.

   a. The curve $y = x^2 + 3$, the $x-$axis and the lines $x = 1$ and $x = 2$ about the $x-$axis. $\left[\textbf{Ans}: \frac{146\pi}{5}\right]$

   b. The curve $y = x^3$, the $y-$axis and the line $y = 1$ about the $y-$axis. $\left[\textbf{Ans}: \frac{3\pi}{5}\right]$

   c. The circle $x^2 + y^2 = 25$ the $x-$axis and thelines $x = 2$ and $x = 3$ about the $x-$axis. $\left[\textbf{Ans}: \frac{56\pi}{3}\right]$

2. Find the volume of the solid generated when the area bounded by thecurve.

   a. $y = x(1-x)$ $\left[\textbf{Ans}: \frac{\pi}{30}\right]$

   b. $y^2 = x(2x-1)^2$ and the $x-$axis is rotated about the $x-$axis. $\left[\textbf{Ans}: \frac{\pi}{48}\right]$

3. Find the volume of the solid formed by revolving the ara enclosed by the curve $y = x^2$, the $y-$axis and the lines $y = 8$, $y = 0$ about the $y-$axis. $\left[\textbf{Ans}: \frac{96\pi}{5}\right]$

4. The region of the circle $x^2 + y^2 = 16$ is revolved about a diameter. Find the volume of the sphere thus generated. $\left[\textbf{Ans}: \frac{256\pi}{3}\right]$

5. Find the volume of the solid of revolution formed when the curve $y = \sin x$ rotates about the $x-$axis between $x = 0$ and $x = \pi$.

   or

   Find the volume generated by revolving about the $x-$axis an arch of the curve $y = \sin x$. $\left[\textbf{Ans}: \frac{\pi^2}{2}\right]$

6. Find the volume the solid obtained by rotating one arch of thecurve $y = \sin 3x$ about the x – axis. $\left[\textbf{Ans}: \frac{\pi^2}{6}\right]$

7. Determine the volume of the solid generated by revolving the area bounded by the parabola $y^2 = 8x$ cut off by its latus - rectum given by $x = 2$ about the

   a. x – axis $[\textbf{Ans}: 16\pi]$

   b. tangent line at the vertex of the parabola. $\left[\textbf{Ans}: \frac{128}{5}\pi\right]$

8. a. Find the volume generated by the revolution of the ellipse $\frac{x^2}{25} + \frac{y^2}{9} = 1$ about x -- axis. $[\textbf{Ans}: 60\pi]$

   b. Find the volume generated by the revolving the ellipse $\frac{x^2}{9} + \frac{y^2}{4} = 1$ about the minor axis. $[\textbf{Ans}: 24\pi]$

   c. The ellipse $\frac{x^2}{9} + \frac{y^2}{4} = 1$ is rotated about the x – axis. Find the the volume of the solid generated. $[\textbf{Ans}: 16\pi]$

9. The ellipse $\frac{x^2}{9} + \frac{y^2}{4} = 1$ is rotated about the x – axis. Find the the volume of the solid generated. $\left[\textbf{Ans}: \frac{320\pi}{3}\right]$

10. Find the volume of the solid generated by the revolution of the area bounded by

    a. $y^2 = 9x$ and $y = 3x$ $\left[\textbf{Ans}: \frac{3\pi}{2}\right]$

    b. $y = x^2$, $y = 2x$ about the x – axis. $\left[\textbf{Ans}: \frac{64\pi}{15}\right]$

11. Find the volume of right circular cone of height h and semi-vertical angle α. $\left[\textbf{Ans}: \frac{\pi}{3}h^3 \tan^2 \alpha\right]$

12. The area between the curves $y^2 = 4ax$ and $x^2 = 4ay$ revolves about $x-$axis. Find the volume of the solid so generated. $\left[\textbf{Ans:} \frac{96a^3}{5}\pi\right]$

13. Find the volume of the solid generated when the region bounded by $y = x^2$ and $y = x$ is revolved about the $y-$axis.

14. Find the volume of the solid generated by the revolution of thearea bounded by the parabola $y^2 = 4x$ and the straight line $4x - 3y + 2 = 0$, about the $y-$axis. $\left[\textbf{Ans:} \frac{\pi}{20}\right]$

15. Find the volume generated by revolving the area bounded by the curve $\sqrt{x} + \sqrt{y} = \sqrt{a}, x = 0\ y = 0$ about

    a. the $x-$axis $\left[\textbf{Ans:} \frac{\pi a^3}{15}\right]$

    b. the $y-$axis. $\left[\textbf{Ans:} \frac{\pi a^3}{15}\right]$

2.7 Mean and R.M.S. Values

2.7.0 Mean (average) value of a function in an interval

If the graph of a function $y = f(x)$ is drawn from $x = a$ to $x = b$ then the average height of the graph above the x-axis is called the Mean value of the function $f(x)$ in $[a, b]$.

$$\text{Mean value of y} = \frac{\text{area under the curve}}{\text{length of the base}}$$

$$= \frac{1}{b-a}\int_a^b y\,dx = \frac{1}{b-a}\int_a^b f(x)dx$$

Current: If a current i varies with time t, then its mean value during the time interval from t_1 to t_2 is given by

$$\frac{1}{(t_2 - t_1)}\int_{t_1}^{t_2} i\,dt$$

2.7.1 Mean square value (M.S.V)

The mean square value of a function $y = f(x)$ over the range $x = a$ to $x = b$ defined as

$$\text{M.S.V.} = \frac{1}{b-a}\int_a^b [f(x)]^2\,dx$$

2.7.2 Method for finding the interval

In the case of trigonometric functions, where the interval is not given, take the interval as equal to one period. The period of the function is the least positive value of the variable by which the variable may be increased or decreased without affecting the value of the function.

Since $\sin x = \sin(2\pi + x)$ the period of $\sin x$ is 2π. Similarly the period of $\sin^{-1}x$, $\cos^{-1}x$, secx and cosecx is 2π. In each of these cases, the interval of integration is $[0, 2\pi]$

Since $\tan x = \tan(\pi + x)$, the period of $\tan x$ is π. Similarly the period of $\cot x$ is π.

i. Consider the function sin$x$ pt

Since $\sin pt = \sin(2\pi + pt)$, the value of $\sin pt$ remains unchanged by increasing pt by 2π. or by increasing t by $2\pi/p$

Hence the period of sin pt is $\dfrac{2\pi}{p}$

Similarly, the period of $\cos pt$ is $\dfrac{2\pi}{p}$

ii. Next, consider the function $\sin^2 pt$. Since $\sin pt = -\sin(\pi + pt)$, it follows that $\sin^2 pt = \sin^2(\pi + pt)$. Hence $\sin^2$ pt remains unaltered by increasing pt by π, or by increasing t by π/p.

$$\therefore \text{ Period of } \sin^2 pt \text{ is } \frac{\pi}{p}$$

iii. Finally, consider the function $\sin pt.\cos qt$, where p and q are integers not having a common factor. The period of sin pt is $2\pi/q$ and that of $\cos qt$ will remain unchanged by increasing t by 2π. Hence the period of sin pt. cosqt is 2π. Similarly the period of sin pt. sin qt is 2p and period of sin pt + cos qt is 2π. Hence the period of $\sin pt$ cos qt and sin $pt + \cos qt$ is 2π.

Ex. The mean value of sin pt is given by

$$\text{Mean value} = \frac{1}{\frac{2\pi}{p} - 0} \int_0^{2\pi/p} \sin pt\, dt = \frac{p}{2\pi}\left(-\frac{\cos pt}{p}\right)_0^{2\pi/p}$$

$$= \frac{1}{2\pi}(\cos 2\pi - \cos 0) = \frac{1}{-2\pi}(1-1) = 0$$

which shows that the mean value of any periodic function over its period is zero. In such cases the root-mean-square value is used.

2.7.3 Root mean square value of a function in an interval

The square root of the mean square value of the function $y = f(x)$ from $x = a$ to $x = b$ $(b > a)$ is called the Root Mean Square value, and is written as R.M.S. value.

$$\text{RMS value} = \sqrt{\frac{1}{b-a}\int_a^b \{f(x)\}^2 dx} = \sqrt{M.S.V}$$

WORKED EXAMPLES

Example 1 Find the mean value of $x^2 e^x$ from $x = 1$ to $x = 5$

Solution

Let $f(x) = x^2 e^x$

$$\text{Mean value} = \frac{1}{5-1}\int_1^5 x^2 e^x dx = \frac{1}{4}\left[e^x(x^2 - 2x + 2)\right]_1^5, \text{ by Bernoull's rule}$$

$$= \frac{1}{4}\left[e^5(25 - 10 + 2) - e(1 - 2 + 2)\right] = \frac{1}{4}(17e^5 - e)$$

$$= \frac{e}{4}(17e^4 - 1)$$

Example 2 Find the mean value of $x^2 - 5x + 6$ between the values of x where the expression vanishes.

Solution

Let $f(x) = x^2 - 5x + 6$. Then $f(x)$ vanishes when $x^2 - 5x + 6 = 0$

i.e., $(x - 2)(x - 3) = 0 \Rightarrow x = 2$ or $x = 3$

$$\text{Mean value of } f(x) = \frac{1}{3-2}\int_2^3 f(x)dx = \int_2^3 (x^2-5x+6)dx$$

$$= \left(\frac{x^3}{3}-\frac{5x^2}{2}+6x\right)_2^3$$

$$= \left(9-\frac{45}{2}+18\right)-\left(\frac{8}{3}-10+12\right)$$

$$= \left(27-\frac{45}{2}\right)-\left(\frac{8}{3}+2\right)$$

$$= \frac{9}{2}-\frac{14}{3}=\frac{27-28}{6}=-\frac{1}{6}$$

Example 3 Find the mean value of the expression $\cos(pt+\alpha).\cos(pt+\beta)$

Solution

We have

$\cos(\pi+pt+\alpha) = -\cos(pt+\alpha)$

and $\cos(\pi+pt+\beta) = -\cos(pt+\beta)$

$\therefore \quad \cos(\pi+pt+\alpha).\cos(\pi+pt+\beta) = \cos(pt+\alpha).\cos(pt+\beta)$

Hence the given function remains unchanged by increasing pt by π, i.e., by increasing t by π/p.

$\therefore$ The period of the given function is π/p and we can, take the range $(0, \pi/p)$

$$\text{Hence its mean value} = \frac{1}{\frac{\pi}{p}-0}\int_0^{\pi/p} \cos(pt+\alpha)\cos(pt+\beta)dt$$

$$= \frac{p}{\pi}\cdot\frac{1}{2}\int_0^{\pi/p} 2\cos(pt+\alpha)\cos(pt+\beta)dt$$

$$= \frac{p}{2\pi}\int_0^{\pi/p} [\cos(2pt+\alpha+\beta)+\cos(\alpha-\beta)]dt$$

$$= \frac{p}{2\pi}\left[\frac{1}{2p}\sin(2pt+\alpha+\beta)+\cos(\alpha-\beta)\cdot t\right]_0^{\pi/p}$$

$$= \frac{p}{2\pi}\left[\frac{1}{2p}\sin(2\pi+\alpha+\beta)+\frac{\pi}{p}\cos(\alpha-\beta)-\frac{1}{2p}\sin(\alpha+\beta)-0\right]$$

$$= \frac{p}{2\pi}\left[\frac{1}{2p}\sin(\alpha+\beta)+\frac{\pi}{p}\cos(\alpha-\beta)-\frac{1}{2p}\sin(\alpha+\beta)\right]$$

$$= \frac{p}{2\pi}\cdot\frac{\pi}{p}\cos(\alpha-\beta)=\frac{1}{2}\cos(\alpha-\beta)$$

Example 4 Find the R.M.S. value of xe^x as x varies from $x = 1$ to $x = 3$

Solution

Let $f(x) = xe^x$

Now M.S. value $= \dfrac{1}{b-a}\displaystyle\int_a^b \{f(x)\}^2 dx = \frac{1}{3-1}\int_1^3 x^2e^{2x}dx$

$$= \frac{1}{2}\left[x^2\frac{e^{2x}}{2}-2x\frac{e^{2x}}{4}+2\frac{e^{2x}}{8}\right], \text{ using Bernoullis rule}$$

$$= \left[\frac{e^{ex}}{2}\left(\frac{x^2}{2}-\frac{x}{2}+\frac{1}{4}\right)\right]_1^3$$

$$= \frac{1}{2}\left[e^6\left(\frac{9}{2}-\frac{3}{2}+\frac{1}{4}\right)-e^2\left(\frac{1}{2}-\frac{1}{2}+\frac{1}{4}\right)\right]$$

$$= \frac{1}{2}\left[e^6\left(3+\frac{1}{4}\right)-e^2\left(\frac{1}{4}\right)\right]=\frac{e^2}{2}\left[e^4\frac{13}{.4}-\frac{1}{4}\right]$$

$$= \frac{e^2}{8}(13e^4-1)$$

R.M.S. value $= \sqrt{\text{M.S. Value}} = \sqrt{\dfrac{e^2}{8}(13e^4-1)} = \dfrac{e}{2\sqrt{2}}\cdot\sqrt{13e^4-1}$

Example 5 Find the R.M.S. value of $\sqrt{\log x}$ over the range $x = 1$ to $x=e$

Solution

Let $f(x) = \sqrt{\log x}$

Now M.S. Value $= \dfrac{1}{e-1}\displaystyle\int_1^e \log x dx = \frac{1}{e-1}[x(\log x-1)]_1^e$, by parts

$$= \frac{1}{e-1}[e(\log e - 1) - (\log 1 - 1)]\frac{1}{e-1}[e(1-1)-(0-1)]$$

$$= \frac{1}{e-1}\text{, R.M.S Value} = \sqrt{\text{M.S. value}} = \frac{1}{\sqrt{e-1}}$$

Example 6 Find the R.M.S. value of $f(t) = a \sin pt + b \cos pt$ as t varies from 0 to 2π where p is an integer.

Solution

$$\text{M.S. Value} = \frac{1}{2\pi - 0}\int_0^{2\pi}(a \sin pt + b \cos pt)^2\,dt$$

$$= \frac{1}{2\pi}\int_0^{2\pi}(a^2 \sin^2 pt + 2ab \sin pt \cos pt + b^2 \cos^2 pt)dt$$

$$= \frac{1}{2\pi}\int_0^{2\pi}(a^2 \sin^2 ptdt + \frac{1}{2\pi}\int_0^{2\pi} 2ab \sin pt \cdot \cos ptdt$$

$$+ \frac{1}{2\pi}\int_0^{2\pi} b^2 \cos^2 pt\,dt$$

$$= \frac{1}{2\pi}\int_0^{2\pi}\frac{1-\cos 2pt}{2}dt + \frac{ab}{2\pi}\int_0^{2\pi}\sin 2ptdt + \frac{b^2}{2\pi}$$

$$+ \int_0^{2\pi}\frac{1+\cos 2pt}{2}$$

$$= \frac{a^2}{4\pi}\left(t - \frac{\sin 2pt}{2p}\right)_0^{2\pi} - \frac{ab}{2\pi}\left(\frac{\cos 2pt}{2p}\right)_0^{2\pi} + \frac{b^2}{4\pi}\left(t + \frac{\sin 2pt}{2p}\right)_0^{2\pi}$$

$$= \frac{a^2}{4\pi}(2\pi - 0) - \frac{ab}{4p\pi}(1-1) + \frac{b^2}{4p\pi}(2\pi + 0)$$

$$[\because \sin 4p\pi = 0 \text{ and } \cos 4p\pi = 1]$$

$$= \frac{a^2}{2} - 0 + \frac{b^2}{2} = \frac{a^2+b^2}{2}$$

$$\text{R.M.S Value} = \sqrt{\text{M.S. value}} = \sqrt{\frac{a^2+b^2}{2}}$$

Example 7 Find the R.M.S. value of the current $i = 2t - 4t^2$ from $t = 0$ to $t = 0.5$ sec.

Solution

Mean square value of

$$i = \frac{1}{0.5-0}\int_0^{0.5}(2t-4t^2)^2\,dt = \frac{1}{0.5}\int_0^{0.5}(4t^2+16t^4-16t^3)\,dt$$

$$= 2\left(4\frac{t^3}{3}+16\frac{t^5}{5}-16\frac{t^4}{4}\right)_0^{1/2}$$

$$= 2\left[\left(\frac{4}{3}\cdot\frac{1}{8}+\frac{16}{5}\cdot\frac{1}{32}-4\cdot\frac{1}{16}\right)-0\right]$$

$$= 2\left[\frac{1}{6}+\frac{1}{10}-\frac{1}{4}\right] = 2\left[\frac{10+6-15}{60}\right] = \frac{1}{30}\text{amp}^2$$

$$\text{R.M.S Value of i} = \frac{1}{\sqrt{30}} = 0.18\,\text{amp}$$

Example 8 Find the R.M.S. value of current over a complete cycle of the current $i = \sin\theta$

Solution

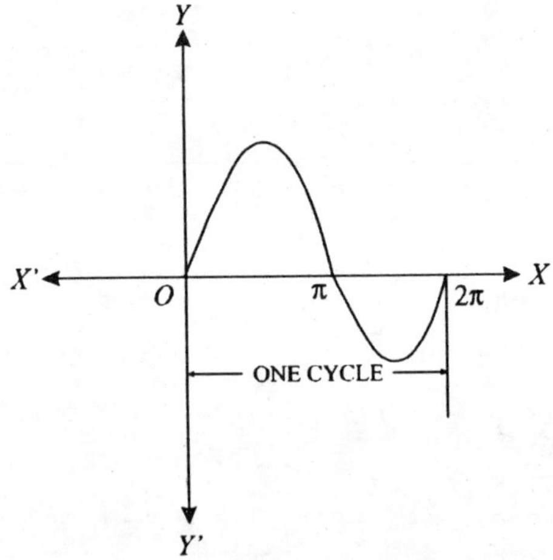

Fig. 2.22

$$\text{Mean square value of } i = \frac{1}{2\pi - 0}\int_0^{2\pi} \sin^2\theta d\theta = \frac{1}{2\pi}\int_0^{2\pi} \frac{1-\cos 2\theta}{2} d\theta$$

$$= \frac{1}{4\pi}\left(\theta - \frac{\sin 2\theta}{2}\right)_0^{2\pi}$$

$$= \frac{1}{4\pi}\left[\left(2\pi - \frac{\sin 4\pi}{2}\right) - (0-0)\right]$$

$$= \frac{1}{4\pi}(2\pi - 0) = \frac{1}{2}$$

$$\text{R.M.S Value of } i = \sqrt{\text{M.S. value}} = \frac{1}{\sqrt{2}}\text{amp}$$

Example 9 An alternating current 'i' at any time t is given by $i = \sin 3t$. Find the R.M.S. value of the current over the range $t = 0$ to $t = \pi/3$.

Solution

$$\text{Mean square value (M.S.V)} = \frac{1}{b-a}\int_a^b [f(x)]^2 dt = \frac{1}{\left(\frac{\pi}{3}\right) - 0}\int_0^{\pi/3} \sin^2 3t\, dt$$

$$= \frac{3}{\pi}\int_0^{\pi/3} \frac{1-\cos 6t}{2} dt = \frac{3}{2\pi}\left(t - \frac{\sin 6t}{6}\right)_0^{\pi/3}$$

$$= \frac{3}{2\pi}\left[\left(\frac{\pi}{3} - \frac{1}{6}\sin 2\pi\right) - (0-0)\right] = \frac{3}{2\pi}\left(\frac{\pi}{3}\right) = \frac{1}{2}$$

$$\text{R.M.S Value} = \sqrt{\text{M.S. value}} = \frac{1}{\sqrt{2}}$$

EXERCISES

I. The mean value

1. Find the mean value of $y^2 = 4x$ from $x = 0$ to $x = 4$ $\left[\textbf{Ans}: \frac{8}{3}\right]$

2. Find the mean value of $x(a-x)$ over the range $0 \le x \le a$ $\left[\textbf{Ans}: \frac{a^2}{6}\right]$

3. Find the mean value of $x^2 - 4x + 3$ between the value of x where the expression vanishes. $\left[\textbf{Ans}: -\frac{2}{3}\right]$

4. Find the mean value of the following function on the given intervals

a. $f(t) = \sin^2 2t$ on $[0, \pi]$ $\left[\textbf{Ans}: \frac{1}{2}\right]$

b. $f(x) = \log x$ on $[1, e]$ $\left[\textbf{Ans}: \frac{1}{e-1}\right]$

c. $f(x) = \sin^3 x$ on $[0, 2\pi]$ $[\textbf{Ans}: 0]$

d. $f(t) = 2\cos t - \sin 3t$ on $\left[0, \frac{\pi}{6}\right]$ $\left[\textbf{Ans}: \frac{4}{\pi}\right]$

5. a. Find the mean value $x^2 e^x$ from $x = 1$ to $x = 5$ $\left[\textbf{Ans}: \frac{1}{4}(17e^5 - e\right]$

b. Find the mean value of $x^2 e^x$ from $x = 1$ to $x = 3$ $\left[\textbf{Ans}: \frac{e}{2}(5e^2 - 1\right]$

6. If the mean value of $\sin x + k \sin 2x$ in the range $x = 0$ to $x = \frac{\pi}{2}$ is zero, find the value of k. $[\textbf{Ans}: -1]$

7. Find the mean value of the expression $\sin(pt + \alpha).\ \sin(pt + \beta)$ over the range $t = 0$ to π/p. $\left[\textbf{Ans}: \frac{1}{2}\cos(\alpha - \beta)\right]$

8. Find the mean value of $\sin^4 pt$ over the range $t = 0$ to $t = \pi/p$.

$\left[\textbf{Ans}: \frac{3}{8}\right]$

9. Find the mean value of the ordinate of the cycloid $x = a(\theta + \sin\theta)$, $y = a(1 - \cos\theta)$ over the range $\theta = -\pi$ to $\theta = \pi$. $\left[\textbf{Ans}: \frac{9}{2}\right]$

II. The root mean square value

10. Find the mean value and R.M.S. value of $y = x^2 + 2$ over the range $(0, 2)$ $\left[\textbf{Ans}: \frac{10}{3}, 3.54\right.$

11. Find the R.M.S. value of $\cos x$ from $x = 0$ to $x = \pi$. $\left[\textbf{Ans}: \frac{1}{\sqrt{2}}\right]$

12. a. Find the R.M.S. value of $\sqrt{8-4x^2}$ between $x = 0$ and $x = 2$

$\left[\textbf{Ans}: \sqrt{\frac{8}{3}}\right]$

b. Find the R.M.S. value of $\sqrt{27-4x^2}$ from $x = 0$ to $x = 3$

$\left[\textbf{Ans}: \sqrt{15}\right]$

13. Determine the root mean square value of the function $y = xe^x$ in the range between $x = 0$ and $x = 1$

$\left[\textbf{Ans}: \frac{\sqrt{e^2-1}}{2}\right]$

14. Find the R.M.S. value of the $\log x$ over the range $x = 1$ to $x = e$

$\left[\textbf{Ans}: \sqrt{\frac{e-2}{e-1}}\right]$

15. Find the R.M.S. value of the expression $(a + b\sin\theta)$ over the interval 0 to 2π.

$\left[\textbf{Ans}: \sqrt{\frac{2a^2+b^2}{2}}\right]$

16. Find the R.M.S. value of the current I = a sinx, over a half wave.

[**Hint :** x varies from 0 to π]

$\left[\textbf{Ans}: \frac{a}{\sqrt{2}}\right]$

17. If the gas confined in a piston expands adiabatically according to the law $pv^n = c$, show that the average value of the pressure as volume increases from v_1 to v_2 is $\frac{p_1v_1 - p_2v_2}{(n-1)(v_2 - v_1)}$

3

Numerical Integration

3.0 Introduction

In this chapter the topic of numerical Analysis is persued further by presenting some numerical methods of integration and solution of first - order ordinary differential equations.

3.1 Numerical Integration

Consider a function $y = f(x)$ and suppose $y_0, y_1, y_2, y_3 ... y_n$ are the specified values of y corresponding values $x_0, x_1, x_2, x_3, ... x_n$ of x with h as the step - length. Suppose we wish to evaluate the integral.

$$I = \int_{x_0}^{x_n} f(x)dx = \int_{x_0}^{x_0+x_n} f(x)dx \qquad ... 1$$

using the specified values $y_0, y_1, y_2, y_3, ... y_n$ of y.
Let us set $x = x_0 + ht$, $dx = h\,dt$ so that $t = 0$ corresponds to $x = x_0$ and $t = n$ corresponds to $x = x_n$.
Then the expression (1) becomes

$$I = \int_0^n f(x_0 + ht)h\,dt \qquad ... 2$$

By employing the Newton's forward interpolation formula, we obtain

$$I = h\int_0^n \left[y_0 + t(\Delta y_0) + \frac{t(t-1)}{2!}(\Delta^2 y_0) + \frac{t(t-1)(t-2)}{3!}(\Delta^3 y_0) + ... \right] dt$$

$$= h\left[y_0 t + \frac{t^2}{2}(\Delta y_0) + \frac{1}{2!}\left(\frac{t^3}{3} - \frac{t^2}{2}\right)(\Delta^2 y_0) + \frac{1}{3!}\left(\frac{t^4}{4} - t^3 + t^2\right)(\Delta^3 y_0) + ... \right]_0^n$$

$$= nh\left[y_0 + \frac{n}{2}(\Delta y_0) + \frac{1}{2!}\left(\frac{n^2}{3} - \frac{n}{2}\right)(\Delta^2 y_0) + \frac{1}{3!}\left(\frac{n^3}{4} - n^2 + n\right)(\Delta^3 y_0) + \ldots\right]_0^n \quad \ldots 3$$

This formula which gives the value of I is known as the **quadrature formula**. By truncating the right hand side of the formula at appropriate stages, we obtain approximate values of I to the desired degrees of accuracy. Below we consider some particular cases of the formula (3).

Trapezoidal Rule

What is the trapezoidal rule?

The trapezoidal rule is based on the Newton-Cotes formula that if one approximates the integrand by an nth order polynomial, then the integral of the function is approximated by the integral of that n^{th} order polynomial. Integrating polynomials is simple and is based on the calculus formula.

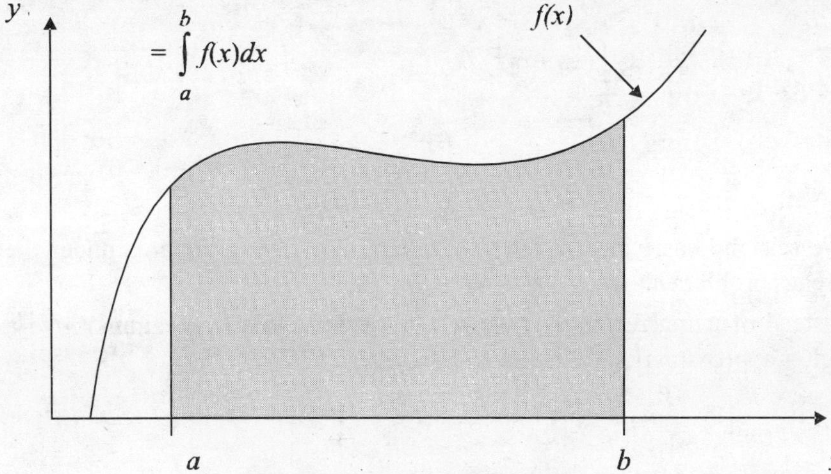

Fig. 3.1: Integration of a function

$$\int_a^b x^n dx = \left(\frac{b^{n+1} - a^{n+1}}{n+1}\right), n \neq -1 \quad \ldots 1$$

So if we want to approximate the integral

$$I = \int_a^b f(x)dx \quad \ldots 2$$

to find the value of the above integral, one assumes

$$f(x) \approx f_n(x) \quad \ldots 3$$

where

$$f_n(x) = a_0 + a_1 x + \ldots + a_{n-1}x^{n-1} + a_n x^n \qquad \ldots 4$$

where $f_n(x)$ is a nth order polynomial. The trapezoidal rule assumes $n = 1$, that is, approximating the integral by a linear polynomial (straight line),

$$\int_a^b f(x)dx \approx \int_a^b f_1(x)dx$$

Derivation of the Trapezoidal Rule

Method 1: Derived from Calculus

$$\int_a^b f(x)dx \approx \int_a^b f_1(x)dx$$

$$= \int_a^b (a_0 + a_1 x)dx$$

$$= a_0(b-a) + a_1\left(\frac{b^2 - a^2}{2}\right) \qquad \ldots 5$$

We saw the basic idea in our first attempt at solving the area under the arches problem above.

Instead of using rectangles, we see that **trapezoids** (trapeziums) give a better approximation to the area.

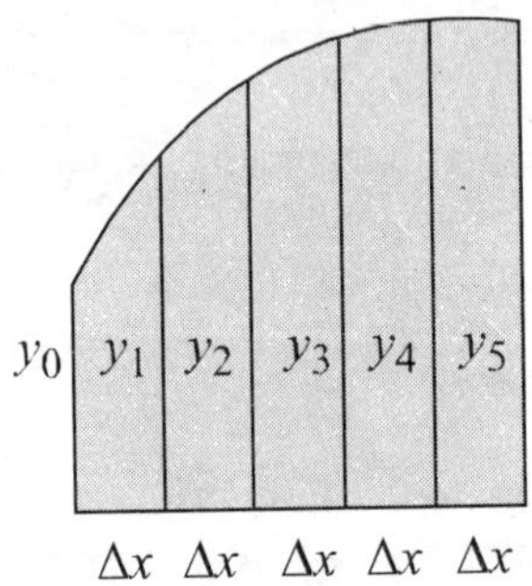

Now, the area of a trapezoid (trapezium) is given by:

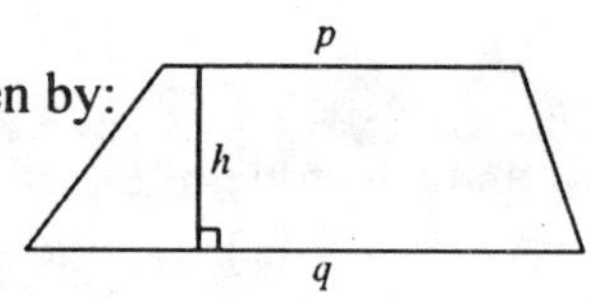

$$\text{Area} = \frac{h}{2}(p+q)$$

So the approximate area under the curve is found by adding the area of the trapezoids. (Our trapezoids are rotated 90° so that their new base is actually the height. So $h = \Delta x$.)

$$\text{Area} \approx \frac{1}{2}(y_0 + y_1)\Delta x + \frac{1}{2}(y_1 + y_2)\Delta x + \frac{1}{2}(y_2 + y_3)\Delta x....$$

We can simplify this to give us the **Trapezoidal Rule**, for n trapezoids:

$$\text{Area} \approx \Delta x\left(\frac{y_0}{2} + y_1 + y_2 + y_3 + ... + \frac{y_n}{2}\right)$$

[This is less calculation than the form used in the text. You may use either.]
To find Δx for the area from $x = a$ to $x = b$, we use:

$$\Delta x = \frac{b-a}{n}$$

and we also need

$$y_0 = f(a)$$
$$y_1 = f(a + \Delta x)$$
$$y_2 = f(a + 2\Delta x)$$
$$y_n = f(b)$$

Note:

We get a better approximation if we take more trapezoids [up to a limit!].
The more trapezoids we take, we have: $\Delta x \to 0$.
We can write (if the curve is above the x-axis only between $x=a$ and $x=b$):

$$\text{Area} = \int_a^b f(x)dx \approx \Delta x\left(\frac{y_0}{2} + y_1 + ... + \frac{y_n}{2}\right)$$

Trapezoids Formula

$$\int_0^n ydx \equiv \int_0^n f(x)dx = \frac{h}{2}\left[(y_0 + y_n) + 2(y_1 + y_2 + y_3 + y_4 + y_5 + \ldots + y_{n-1})\right]$$

WORKED EXAMPLES

Example 1 Evaluate $\int_1^7 f(x)dx$, using the trapezoidal rule, given

x	1	2	3	4	5	6	7
y	2.105	2.808	3.614	4.604	5.857	7451	9.467

Here the step-length $h = 1$. Also $x_0 = 1, x_1 = 2, ..., x_e = x_n = 7$
$y = 2.105, y_1 = 2.808, ... y_6 = y_n = 9.467$
Hence the Trapezoidal rule gives

$$\int_1^7 f(x)dx = \frac{1}{2}h[(y_0+y_6)+2(y_1+y_2+...+y_5)]$$

$$\frac{1}{2}\times 1\times[(2.105+9.467)+2\times(2.808+3.614+4.604+5.857+7.451)]$$

$$= 30.12$$

Example 2 Evaluate $\int_0^6 \frac{dx}{1+x^2}$ by using the trapezoidal rule.

Solution

Here, $f(x)=\frac{dx}{1+x^2}$, $x_0=0$, $x_n=6$

Let us divide the interval $[x_0, x_n] = [0.6]$ into six equal parts with $h = 1$ as the length of each part, so that $n = 6$. Let us find the values of $y = f(x)$ at $x_0=0$, $x_1 = 1$, $x_2 = 2$, $x_3 = 3$ and $x_n = x_6$

These values are shown in the following table.

x	$x_0 = 0$	$x_1 = 1$	$x_2 = 2$	$x_3 = 3$	$x_4 = 4$	$x_5 = 5$	$x_6 = 6$
$y = f(x)$	$y_0 = 1$	$y = \frac{1}{2}$	$y_2 = \frac{1}{5}$	$y_3 = \frac{1}{10}$	$y_4 = \frac{1}{17}$	$y_5 = \frac{1}{26}$	$y_6 = \frac{1}{37}$

Now, the Trapezoidal rule gives

$$\int_1^b \frac{dx}{1+x^2} = \frac{1}{2}h[(y_0+y_6)+2(y_1+y_2+y_3+y_4+y_5)]$$

$$= \frac{1}{2}\left[\left(1+\frac{1}{37}\right)+2\left(\frac{1}{2}+\frac{1}{5}+\frac{1}{10}+\frac{1}{17}+\frac{1}{26}\right)\right]$$

$$= \frac{1}{2}[1.027027+2\times(0.897285)] = 1.4108$$

Remarks: We note that the exact value of the given integral is $\tan^{-1} 6 = 1.405648$ in radians. The value of the integral obtained by the use of the trapezoidal rule (as worked out above) agrees with the exact value up to two places of decimals.

Example 3 Evaluate $\int_{1}^{0} \frac{dx}{1+x}$ by the trapezaidal rule by considering eight sub-intervals of the interval [0, 1]. Hence find an approximate value of log2.

Solution

Here, the interval [0, 1] has to be divided into eight sub-intervals.

Accordingly, we take $h=\frac{1}{8}, x_0=0,\ x8=1$

We compute the values of $y=\frac{1}{1+x}$ corresponding to the values

$x_0=0, x_1=\frac{1}{8}, x_2=\frac{1}{4}, ...x_7=\frac{7}{8}, x_8=1$ These values are shown in the following Table.

x :	0	0.125	0.25	0.375	0.5	0.625	0.75	0.875	1.0
y :	1	0.8889	0.8	0.7273	0.6667	0.6154	0.5714	0.5333	0.5

$$\int_0^1 \frac{dx}{1+x} = \frac{1}{2}h[(y_0+y_6)+2(y_1+y_2+y_3+y_4+y_5)]$$

$$= \frac{1}{2}\times\frac{1}{8}\times(1.5+2\times 4.803)=0.694125 \quad \text{... i}$$

The exact value of the integral is

$$\int_0^1 \frac{dx}{1+x} = [\log(1+x)]_0^1 = \log 2 \quad \text{... ii}$$

From (i) and (ii), we find that log2 = 0.694125

Example 4 Employing the trapezoidal rule, evaluate $\int_0^{\pi/2}$ sinx dx by dividing the interval [0, π/2] to ten equal parts.

Solution

Here, the interval [0,π/2] = [0,90] has to be divided into ten equal sub-intervals. Accordingly, we take $h = (1/10)\ (\pi/2) = 9°$, $x_0 = 0$ and $x_{10} = 90°$ and evaluate $y = f(x) = \sin x$ for $x_0 = 0$, $x_1 = 9°$, $x_2 = 18°, ...x_{10} = 90°$

The values are shown in the following table.

x	0°	9°	18°	27°	36°	45°	54°	63°	72°	81°	90°
y	0	0.1564	0.309	0.454	0.5878	0.7071	0.809	0.891	0.9511	0.9877	1.0000

The trapezoidal rule now gives

$$\int_0^{\pi/2} \sin x\,dx = \frac{1}{2}h[(y_0+y_{10})+2(y_1+y_2+y_3+y_4+y_5+y_6+y_7+y_8+y_9]$$

$$= \frac{1}{2}\times\frac{\pi}{20}\times[(0+1.0000)+2(0.1564+0.309+0.454+$$
$$+0.5878+0.7071+0.809+0.891+0.9511+0.9877)]$$

$$= \frac{\pi}{40}[1.0000+(2\times 5.8531)] = (0.317655\pi)$$

$$= 0.9983 \text{ (tanking } \pi = 22/7)$$

Note: The exact value of the given integral is cos 0 =1

Example 5 Using the trapezoidal rule, evaluate $\int_0^1 e^{-x}$ by dividing the interval [0, 1] into 10 equal parts. Compare with the exact result.

Solution

We divide the interval [0, 1] into ten equal parts by taking $h = 0.1$ we compute the values of $y = f(x) = e^{-x}$ for $x = 0$, $x_1 = 0.1$, $x_2 = 0.2, \ldots x_{10} = 1$ Theses value are shown below:

x	0	0.1	0.2	0.3	0.4	0.5	0.6	0.7	0.8	0.9	1.0
$f(x)$	1	0.9048	0.8187	0.7408	0.6703	0.6065	0.5488	0.4966	0.4493	0.4066	0.3679

The trapezoidal rule now gives

$$\int_0^1 e^{-x}dx = \frac{1}{2}h[(y_0+y_{10})+2(y_1+y_2+y_3+y_4+y_5+y_6+y_7+y_8+y_9]$$

$$= \frac{1}{2}\times 0.1\times[(1+0.3679)+2(0.9048+0.8187+0.7408$$
$$0.6703+0.6065+05488+0.4966+0.4493+0.4066)]$$

$$= 0.632635 \qquad \ldots \text{ i}$$

The exact value of the integral is

$$\int_0^1 e^{-x}dx = [-e^{-x}]_0^1 = 1-0.3679 = 0.63212 \qquad \ldots \text{ ii}$$

Comparing (i) and (ii) we find that the result obtained with the aid of the trapezoidal rule agrees with the exact result up to second place of the decimal.

Example 6 A plane area is bounded by a curve, the x - axis and two extreme ordinates. The area is divided into five parts by equidistant ordinates 2cms apart, the heights of the ordinates being 21.65, 21.04, 20.35, 19.61, 18.75 and 17.80 cms respectively. Find the approximate value of the area.

Solution

Let us take first ordinate at x_0 the second at $x_1 = x_0 + 2$the third at x_2=x_1=2 and so on, and the last at $x_5 = x_4 + 2$. Then, if at $y = f(x)$ is given curve, we have, by what is given,

$y_0 = 21.65,\ y_1 = 21.04,\ y_2 = 20.35,\ y_3 = 19.61,\ y_4 = 1875,\ y_5 = 17.80$

The required area is specified by the integral $\int_{x_0}^{x_5} f(x) = dx$

$$= \frac{1}{2}h[(y_0 + y_5) + 2(y_1 + y_2 + y_3 + y_4)]$$

$$= \frac{1}{2} \times 2 \times [(21.65 + 17.80) + 2 \times (21.04 + 20.35 + 19.61 + 18.75)]$$

$$= 198.95 \text{ sq.cms}$$

EXERCISES

1. The following table gives the values of a functions $y = f(x)$ for some equidistant values of x

x	0.0	0.25	0.05	0.75	1.0
$f(x)$	0.79788	0.77334	0.70413	0.60227	0.48394

Evaluate $\int_0^1 f(x)dx$ by using the trapezoidal rule.

[Ans : 0.68016]

2. Evaluate $\int_4^{5.2} \log x\, dx$ by using the trapezoidal rule, given

x	4.0	4.2	4.4	4.6	4.8	5.0	5.2
log$x$	1.3863	1.4351	1.4816	1.5261	1.5686	1.6094	1.6487

[Ans :1.827655]

3. Evaluate the following integrals by using the trapezoidal rule:

i. $\int_0^1 x^3 dx$ with $h = 0.2$, ii. $\int_{-3}^{3} x^4 dx$ with $h = 1$

[Ans : (i) 0.26, (ii) 115]

4. By using the trapezoidal rule, evaluate $\int_0^{\frac{\pi}{2}} \cos x \, dx$, Divided the interval to 10 equal parts. **[Ans :0.998]**

5. Using the trapezoidal rule, evaluate $\int_0^1 e^{-x^2}$, dividing the interval into ten equal parts. **[Ans : 0.746211]**

6. By using the trapezoidal rule, find the area below the curve $y = f(x)$ in the interval [7.47, 7.52] given

x	7.47	7.48	7.49	7.50	7.51	7.52
y	1.93	1.95	1.98	2.01	0.03	2.06

[Ans :0.9965]

3.2 Simpson's One-third Rule

In the quadrature formula (3) of Section 3.1, suppose we take $n = 2$ and neglect terms containing $\Delta^3 y_0$, $\Delta^4 y_0$..... Then we get
Using the results

$$\Delta y_0 = y_1 - y_0$$

$$\Delta^2 y_0 = \Delta y_1 - \Delta y_0 = (y_2 - y_1) - (y_1 - y_0) = y_2 - 2y_1 + y_0$$

this becomes

$$\int_{x_0}^{x_2} f(x)dx = \frac{1}{3}h(y_0 + 4y_1 + y_2)$$

Similarly, we obtain

$$\int_{x_2}^{x_4} f(x)dx = \frac{1}{3}h(y_2 + 4y_3 + y_4)$$

....

$$\int_{x_{n-2}}^{x_n} f(x) = \frac{1}{3}h(y_{n-2} + 4y_{n-1} + y_2)$$

(This is valid only when is even) Adding all integrals obtained above, we obtain

$$\int_{x_0}^{x_n} f(x)dx \equiv \int_{x_0}^{x_0+nh} f(x)dx$$

$$= \frac{1}{3}h\left[(y_0 + y_n) + 4(y_1 + y_3 + ... + y_{n-1}) + 2(y_2 + y_4 + ... + y_{n-2})\right]$$

This formula is known as the ***Simpson's 1/3rd Rule,*** or, simply, the ***Simpson's Rule*** (named after the english mathematician T. Simpson (1710-1761) It is to be emphasized that this rule is applicable only when ***n is an even number.***

WORKED EXAMPLES

Example 1 By using the Simpson's 1/3 Rule, evaluate given $\int_1^5 f(x)dx$

x:	1	2	3	4	5
f(x)	13	50	70	80	100

Solution

In the given Table, the interval [1, 5] is divided into four equal parts, each of length $h = 1$

Also, $x_0 = 1, x_1 = 2, x_2 = 3, x_3 = 4, x_4 = 5$ and the corresponding values of $y = f(x)$ are $y_0 = 13, y_1 = 50, y_2 = 70, y_3 = 80, y_4 = 100$ Hence the ***Simpson's 1/3 Rule*** gives

$$\int_1^5 f(x)dx = \int_{x_0}^{x_4} f(x)dx = \frac{1}{3}h\left[(y_0 + y_2) + 4(y_1 + y_3) + 2y_2\right]$$

$$= \frac{1}{3} \times 1 \times \left[(13 + 100) = 4(50 + 80) + 2 \times 70\right]$$

$$= 257.6667$$

Example 2 By using the Simpson's 1/3, Rule, evaluate $\int_0^1 \frac{dx}{1+x^2}$ dividing the interval into six equal parts. Hence find an approximate value of π .

Solution

Here, the interval [0, 1] is to be divided into six equal parts. Hence, we take h = 1/6 and compute the values of $y = \frac{1}{1+x^2}$ at the points $x_0 = 0$, $x_1 = 1/6$, $x_2 = 1/3$, ... $x_6 = 1$; the values are shown in the following table.

x	$x_0 = 0$	$x_1 = 1/6$	$x_2 = 1/3$	$x_3 = 1/2$	$x_4 = 2/3$	$x_5 = 5/6$	$x_6 = 1$
$f(x)$	$y_0 = 1$	$y_1 = 0.97297$	$y_2 = 0.9$	$y_3 = 0.8$	$y_4 = 0.69231$	$y_5 = 0.59016$	$y_6 = 0.5$

The Simpson's 1/3 Rule now gives

$$\int_0^1 \frac{dx}{1+x^2} = \int_{x_0}^{x_6} f(x)dx = \frac{1}{3}h[y_0 + y_6) + 4(y_1 + y_3 + y_5) + 2(y_2 + y_4)]$$

$$= \frac{1}{3}\times\frac{1}{6}\times[(1+0.5) + 4(0.97297 + 0.8 + .59016)$$

$$+2(0.9 + 0.69231)]$$

$$= \frac{1}{3}\times\frac{1}{6}\times[1.5 + 4\times 2.36313 + 2\times 1.59231]$$

$$= 0.785397 \qquad \ldots \text{ i}$$

The exact value of the integral is

$$\int_0^1 \frac{dx}{1+x^2} = \left[\tan^{-1} x\right]_0^1 = \frac{\pi}{4} \qquad \ldots \text{ ii}$$

From (i) and (ii) we get

$$\pi = 4 \times 0.785397 = 3.141588$$

Example 3 Evaluate $\int_0^1 \frac{xdx}{1+x^2}$ by using Simpon's 1/3 rule taking six equal strips, and hence deduce an approximate value of log2.

Solution

Here, the interval [0,1] is to be divided into six equal parts. Hence, we take $h = 1/6$ and compute the values of $y = x/(1+x^2)$at the points $x_0 = 0$, $x_1 = 1/6$, $x_2 = 2/3 \ldots x_6 = 1$; The values are shown in the following table.

x	$x_0=0$	$x_1=1/6$	$x_2=1/3$	$x_3=1/2$	$x_4=2/3$	$x_5=5/6$	$x_6=1$
$f(x)$	$y_0=0$	$y_1=0.16216$	$y_2 = 0.3$	$y_3 = 0.4$	$y_4 = 0.46154$	$y_5 = 0.4918$	$y_6=0.5$

The simpson's 1/3 rule now gives

$$\int_0^1 \frac{xdx}{1+x^2} = \frac{1}{3}h[(y_0 + y_6) + 4(y_1 + y_3 + y_5) + 2(y_2 + y_4)]$$

$$= \frac{1}{3} \times \frac{1}{6} \times [(0+0.5)+4(0.16216+0.4+0.4918)$$

$$+2(0.3+0.46154)]$$

$$= \frac{1}{18}[0.5+4.21584+1.52308]$$

$$= 0.34661 \quad \ldots \text{ i}$$

Next, by direct integration, we find

$$\int_0^1 \frac{xdx}{1+x^2} = \frac{1}{2}\left[\log(1+x^2)\right]_0^1 = \frac{1}{2}\log 2 \quad \ldots \text{ ii}$$

From (i) and (ii) we get

$$\log 2 = 2 \times 0.34661 = 0.69322$$

Example 4 Evaluate $\int_0^{\pi/2} \sqrt{\cos\theta}\ d\theta$ by dividing the interval into eight equal parts.

Solution

Here, we take $h = \frac{1}{8}\left(\frac{\pi}{2} - 0\right) = \frac{\pi}{16}$ and compute the values of $\varphi = f(x) = \sqrt{\cos\theta}$ for θ values are shown in the following tables.

θ	0	π/16	π/8	3π/16	π/4	5π/16	3π/8	7π/16	π/2
φ=*f*(θ)	1	0.9903	0.9612	0.9118	0.8409	0.7454	0.6186	0.4417	0

The Simpson's 1/3 rule now gives

$$\int_0^{\pi/2} \sqrt{\cos\theta}d\theta = \int_{\theta_0}^{\theta_8} f(\theta) = d\theta = \frac{1}{3}h\left[(\varphi_0+\varphi_8)+4(\varphi_1+\varphi_3+\varphi_5+\varphi_7)\right.$$

$$\left.+2(\varphi_2+\varphi_4+\varphi_6)\right]$$

$$= \frac{1}{3} \times \frac{\pi}{16} \times [(1+0)+4(0.9903+0.9118+0.7454$$

$$0.4417)+2(0.9612+0.8409.+0.6186)]$$

$$= \frac{1}{3} \times \frac{\pi}{16} \times [1 + (4 \times 3.0892) + (2 \times 2.4207)]$$

$$= (0.37913)\pi = 1.19155$$

Example 5 Evaluate $\int_0^5 \frac{dx}{4x+5}$ by using the Simpson's 1/3 Rule with the aid of 11 ordinates. Hence find log 5.

Solution

Here, we have to use 11 ordinates. This means that we have to divide the interval [0, 5] into ten equal sub-intervals. Accordingly, we have $h = 1/2$. Also, $x_0 = 0,\ x_1 = 0.5, x_2 = 1.0, \ldots.\ x_{10} = 5$ We compute the corresponding values of $y = f(x) = \frac{1}{4x+5}$ and the values are shown in the following table.

x	y
$x_0 = 0$	$y_0 = 0.2$
$x_1 = 0.5$	$y_1 = 0.14286$
$x_2 = 1.0$	$y_2 = 0.11111$
$x_3 = 1.5$	$y_3 = 0.09091$
$x_4 = 2.0$	$y_4 = 0.09692$
$x_5 = 2.5$	$y_5 = 0.06667$
$x_6 = 3.0$	$y_6 = 0.05882$
$x_7 = 3.5$	$y_7 = 0.05263$
$x_8 = 4.0$	$y_8 = 0.04762$
$x_9 = 4.5$	$y_9 = 0.04348$
$x_{10} = 5.0$	$y_{10} = 0.04$

Now, the Simpson's 1/3 Rule gives

$$\int_0^5 \frac{dx}{4x+5} = \frac{1}{3}h[(y_0 + y_{10}) + 4(y_1 + y_3 + y_5 + y_7 + y_9$$

$$+ 2(y_2 + y_4 + y_6 + y_8)]$$

$$= \frac{1}{3} \times \frac{1}{2}[(0.2 + 0.04) + 4(0.14286 + 0.09091 + 0.0667$$

$$+0.05263+0.04348)+2(0.11111+0.07692+0.05882+0.04762)]$$

$$= \frac{1}{6}\times[0.24+1.5862+0.58894]=0.40252 \qquad \dots \text{i}$$

Next, by direct integration, we get

$$\int_0^5 \frac{dx}{4x+5} = \frac{1}{4}[\log 4x+5]_0^5 = \frac{1}{4}[\log 25-\log 5]=\frac{1}{4}\log 5 \qquad \dots \text{ii}$$

From (i) and (ii) we get

$$\log 5 = 4\times(0.40252)=1.6101$$

Example 6 By using Simpson's 1/3 Rule, evaluate $\int_0^1 e^{-x^2}$ taking $h = 0.1$

Solution

We first tabulate the values of $y = f(x) = e^{-x^2}$ for $x_0 = 0, x_1 = 0.1, x_2=0.2\dots$ $x_{10} = 1$; the computations are summarized in the Table shown below.

x	x^2	$y=f(x)=e^{-x^2}$
$x_0 = 0$	0	$y_0 = 1$
$x_1 = 0.1$	0.01	$y_1 = 0.99005$
$x_2 = 0.2$	0.04	$y_2 = 0.96080$
$x_3 = 0.3$	0.09	$y_3 = 0.91393$
$x_4 = 0.4$	0.16	$y_4 = 0.85214$
$x_5 = 0.5$	0.25	$y_5 = 0.77880$
$x_6 = 0.6$	0.36	$y_6 = 0.96768$
$x_7 = 0.7$	0.49	$y_7 = 0.61263$
$x_8 = 0.8$	0.64	$y_8 = 0.52730$
$x_9 = 0.9$	0.81	$y_9 = 0.44490$
$x_{10} = 0.10$	1.00	$y_{10} = 0.36788$

Using the entries in the Table, the Simpson's 1/3 rule gives

$$\int_0^1 e^{-x^2}dx = \frac{0.1}{3}\times[(1+0.36788)+4(0.99005+0.91393+0.77880+$$

$$0.61263+0.44490)+2(0.96080+0.85214+0.69768+0.52730)]$$

$$= \frac{0.1}{3}\times[1.36788+4\times 3.74031+2\times 3.03792]$$

$$= 0.746832$$

Example7 A solid of revolution is formed by rotating about the x - axis the area between the x-axis, the lines $x = 0$ and $x = 1$ and a curve through the points with the following coordinates.

x	0.00	0.25	0.50	0.75	1.00
y	10000	0.9896	0.9589	0.9089	0.8415

Using the Simpson's rule, find the volume of the solid formed.

Solution

We first note that the required volume is given by

$$v = \pi \int_0^1 y^2 dx$$

We now evaluate the integral by first computing the values $Y = y^2$ of for the specified values x of using the given table. The values are shown in the following Table:

x	x_0=0.00	x_1=0.25	x_2=0.50	x_3=0.75	x_4=1.00
Y=y^2	Y_0=1	Y_1=0.97931	Y_2=0.91949	Y_3=0.8261	Y_4=0.70812

Now, the Simpson's rule gives the required volume as

$$V = \pi \times \frac{1}{3} h[(Y_0 + Y_4) + 4(Y_1 + Y_3) + 2Y_2]$$

$$= \frac{\pi}{3} \times (0.25) \times [(1 + 0.70812) + 4(0.97931 + 0.8261) + 2(0.91949)]$$

$$= \frac{\pi}{3} \times (0.25) \times [1.70812 + 7.22164 + 1.83898] = (0.897395)\pi$$

Example 8 A rocket is launched from the ground. Its acceleration f is registered during the first 80 seconds and is tabulated below. Using the Simpson's rule, find the velocity of the rocket $t = 80$ at seconds:

t(sec)	0	10	20	30	40	50	60	70	80
f(cm/sec^2)	30	31.63	33.34	35.47	37.75	40.33	43.25	46.69	50.67

Solution If v is the velocity at time t, we have $\frac{dv}{dt} = f$ so that $v = \int_0^t f\, dt$.

The velocity at $t = 80$ is therefore given by $v = \int_0^{80} f\, dt$

From the Table we note that $h = 10$ and $n = 8$. Therefore the Simpson's rule gives

$$v = \int_0^{80} f\,dt = \frac{1}{3} \times 10 \times [(30 + 50.67) + 4(31.63 + 35.47 + 40.33 + 46.69) + 2(33.34 + 37.75 + 43.25)]$$

Example 7 Using $n = 5$, approximate $\int_0^1 \sqrt{x^2+1}\,dx$

Solution

Here, $a = 0$ and $b = 1$.

$$\Delta x = \frac{b-a}{n} = \frac{1-0}{5} = 0.2$$

$$y_0 = f(a) = f(0) = \sqrt{0^2+1} = 1$$

$$y_1 = f(a+\Delta x) = f(0.2) = \sqrt{0.2^2+1} = 1.0198039$$

$$y_2 = f(a+2\Delta x) = f(0.4) = \sqrt{0.4^2+1} = 1.0770330$$

$$y_3 = f(a+3\Delta x) = f(0.6) = \sqrt{0.6^2+1} = 1.1661904$$

$$y_4 = f(a+4\Delta x) = f(0.8) = \sqrt{0.8^2+1} = 1.2806248$$

$$y_5 = f(b) = f(1) = \sqrt{1^2+1} = 1.1412136$$

So the area ≈

$$0.2\left(\frac{1}{2} \times 1 + 1.0198039 + 1.0770330 + 1.1661904 + 1.2806248 ... + \frac{1}{2} \times 1.4142136\right)$$

$$= 1.150$$

So $\int_0^1 \sqrt{x^2+1}\,dx \approx 1.150$

Exercise 8 Using $n = 5$, approximate $\int_0^1 \sqrt{x^2+1}\,dx$

Solution

Here, $a = 0$ and $b = 1$.

$$\Delta x = \frac{b-a}{n} = \frac{1-0}{5} = 0.2$$

$$\Delta x = \frac{b-a}{n} = \frac{1-0}{5} = 0.2$$

$$y_0 = f(a) = f(0) = \sqrt{0^2+1} = 1$$

$$y_1 = f(a+\Delta x) = f(0.2) = \sqrt{0.2^2+1} = 1.0198039$$

$$y_2 = f(a+2\Delta x) = f(0.4) = \sqrt{0.4^2+1} = 1.0770330$$

$$y_3 = f(a+3\Delta x) = f(0.6) = \sqrt{0.6^2+1} = 1.1661904$$

$$y_4 = f(a+4\Delta x) = f(0.8) = \sqrt{0.8^2+1} = 1.2806248$$

$$y_5 = f(b) = f(1) = \sqrt{1^2+1} = 1.1412136$$

So the area ≈

$$0.2\left(\frac{1}{2}\times 1 + 1.0198039 + 1.0770330 + 1.1661904 + 1.2806248 ... + \frac{1}{2}\times 1.4142136\right)$$

$$= 1.150$$

Let us take $n = 1$ in the quadrature formula (3) and ignoring terms consisting $\Delta^2 y0$, $\Delta^3 y0$, ...

Then we get

$$\int_{x_0}^{x_1} f(x)dx = h\left[y_0 + \frac{1}{2}(\Delta y_0)\right] = h\left[y_0 + \frac{1}{2}(y_1 - y_0)\right] = \frac{h}{2}[y_0 + y_1]$$

Similarly, we get

$$\int_{x_1}^{x_2} f(x)dx = \frac{h}{2}[y_1 + y_2]$$

$$\int_{x_2}^{x_3} f(x)dx = \frac{h}{2}[y_2 + y_3]$$

$$\int_{x_{n-1}}^{x_n} f(x)dx = \frac{h}{2}[y_{n-1} + y_n]$$

Adding all the n integrals found above , we get

$$I = \int_{x_0}^{x_n} f(x)dx \equiv \int_{x_0}^{x_0+n_n} f(x)dx$$

$$= \frac{h}{2}[(y_0 + y_n) + 2(y_1 + y_2 + y_3 + y_{n-1})]$$

This formula is known as the Trapezoidal rule for numerical integration. Geometrically, this rule states that the area under approximately equal to the sum of the areas of the trapeziums with vertices at

$(x_r, 0), (x_{r+1}, 0), (x_{r+1}, y_{r+1}), (x_r, y_r), \quad r = 0, 1, 2, ... n$

EXERCISES

1. Using the Simpson 's 1/3 rule, evaluate $\int_0^2 f(x)dx$ given

x	0.0	0.05	1.0	1.5	2.0
$f(x)$	0.399	0.352	0.242	0.129	0.054

[Ans : 0.477]

2. Evaluate $\int_0^4 e^x\,dx$ Simpson 's 1/3 rule, given that $e = 2.72$, $e^2 = 7.39$ $e^3 = 20.09$, $e^4 = 54.6$ **[Ans :53.87]**

3. The following table gives the values of $y = f(x)$

x	0	1	2	3	4	5	6
$f(x)$	0.146	0.161	0.176	0.190	0.204	0.217	0.230

Evaluate $\int_0^6 f(x)dx$ by using the Simpson 's 1/3 rule.

[Ans : 1.13625]

4. Evaluate $\int_4^5 \log x\,dx$ by using the Simpson's 1/3 rule with the aid of the data given in Exercise 2 of section 5.2

[Ans :1.827847]

5. Evaluate the following integrals by using the Simpson's 1/3 rule.

a. $\int_0^6 \frac{dx}{1+x^2}$ with h = 1,

b. $\int_0^1 \frac{dx}{1+x}$ with $h = 1/8$,

c. $\int_0^1 \frac{dx}{1+x^2}$ with $h = 0.25$

d. $\int_0^2 \frac{dx}{1+x^3}$ with $n = 4$

e. $\int_0^1 \frac{x^2}{1+x^3}$ with n = 4 f. $\int_1^5 \log x \, dx$ with n = 10

[**Ans :** (a) 1.3662, (b) 0.6932 (c) 0.7854, (d) 1.096, (e) 0.23108; 0.69324, (f) 177.483, (g) 0.9985 (h) 4.0467]

6. Evaluate $\int_2^8 \frac{dx}{\log_{10} x}$ by using the Simpson's 1/3 rule with aid of seven ordinates (i.e. six sub-intervals.) [**Ans :** 9.7203]
7. Use Simpson's 1/3 rule with $h = 0.125$ to show that

   $$\int_2^8 e^{-x^2/2} dx = 0.85581$$
8. Use Simpson's 1/3 rule with $h = 0.1$ with to show that

   $$\int_0^1 \frac{\log(1+x^2)}{1+x^2} dx = 0.1730$$
9. Use Simpson's 1/3 rule, show that $\int_0^{\pi/2} \sqrt{\sin x}\, dx = \begin{cases} 1.17823 & \text{for } n=4 \\ 1.18728 & \text{for } n=6 \end{cases}$
10. By Simpson's 1/3 rule, find the area below the curve $y = f(x)$ in the interval [1, 4] given

x	1	1.5	2	2.5	3	3.5	4
y	2	2.4	2.7	2.8	3	2.6	2.1

[**Ans :** 7.78]

4

Differential Equations

4.0 Introduction

An equation involving a function and its derivatives is called a differential equation. A large number of problems in applied sciences and engineering ultimately depend on the solution of differential equations. Therefore Differential Equations play a significant role in applie sciences.

In this chapter we study the formation and the solution of differential equations of first order and first degree.

4.1 Differential Equations

An equation involving a function of one (or more) independent variable and its derivative is called a differential equation. A differential equation is said to be ordinary differential equation if the derivatives appearing in it are ordinary such as $\frac{dy}{dx}, \frac{d^2y}{dx^2}$ etc.,

For examples:

$$\frac{dy}{dx} = \frac{\sqrt{1-y^2}}{\sqrt{1-x^2}} \qquad \dots 1$$

$$x^2\frac{d^2y}{dx^2} - 2x\frac{dy}{dx} + 2y = 0 \qquad \dots 2$$

$$x\left(\frac{dy}{dx}\right)^2 - y\frac{dy}{dx} + 5 = 0 \qquad \dots 3$$

$$y = \frac{dy}{dx} + \sqrt{1+\left(\frac{dy}{dx}\right)^4} \qquad \dots 4$$

$$\left[1+\left(\frac{dy}{dx}\right)^2\right]^{\frac{3}{2}} = \frac{d^2y}{dx^2} \qquad \dots 5$$

Are some examples of ordinary differential equations.

4.2 Order and Degree of a Differential Equations

The order of a differential equation is the order of the highest order derivative appearing in it.

The degree of a differential equation is the highest degree of the highest order derivative appearing in it when the differential coefficients are free from radicals and fractions i.e., the differential equation is written as a rational integral equation in the highest order derivative.

In the above examples (1) is of first order and first degree(2) is of order 2 and degree 1 and (3) is of order 1 and degree 2. To find the order and degree of (4), we must write it in the form :

$$y - x\frac{dy}{dx} = \sqrt{1+\left(\frac{dy}{dx}\right)^4}$$

Squaring, we get, $\left(y - x\frac{dy}{dx}\right)^2 = 1+\left(\frac{dy}{dx}\right)^4$

Now this equation is free from radical signs and derivatives do not appear in the denominator. Therefore this is a differential equation of order 1 and degree 4.

Similarly squaring (5), we get,

$$y - x\frac{dy}{dx} = \sqrt{1+\left(\frac{dy}{dx}\right)^4}$$

$$\left[1+\left(\frac{dy}{dx}\right)^2\right]^3 = \left(\frac{d^2y}{dx^2}\right)^2$$

Hence this equation is of the second order and second degree.

4.3 Formation of Differential Equations

Consider the equation $y^2 = 4ax$ ... 1

We know that (1) represents a parabola with vertex at the origin and x – *axis* as the axis. For different values of a we get different parabolas having their vertices at the origin and x - *axis* as the axis. Hence (1) represents a family of parabolas. Here is called the parameter.

Now differentiating (1) w.r.t 'x' we get $2y\frac{dy}{dx} = 4a$ Substituting the value of $4a$ in (1), we get $y^2 = 2y\frac{dy}{dx}x$ ... 2

This is the differential equation obtained from (1) by eliminating the arbitrary constant a. Hence the differential equation represents a family of parabolas given by (1)

In general suppose the equation $F(x, y, a) = 0$ ... 3

represents a family of curves in which a is the parameter. Differentiating (3) w.r.t. x and eliminating we get an equation of the form

$$\cos A \cos B = \frac{1}{2}[\cos(A+B)+\cos(A-B)]$$

$$\phi\left(x, y, \frac{dy}{dx}\right) = 0 \quad \ldots 4$$

Which is a differential equation of first order.

Further the differential equation (4) represents a family of curves given by (3). If there are two constants (parameters in a relation then in order to eliminate these constants we have to differentiate twice. Hence the differential equations of order 2 are formed and so on.

4.4 Solution of Differential Equations

In the above discussion the differential equation (2) is formed from (1) by eliminating the arbitrary constant Hence equation (1) must satisfy (2) Therefore (1) is called a solution of (2). Since (2) is satisfied by (1) for any value of a, (1) is called the general solution of (2) Also the differential equation (2) is satisfied for a particular value of a. For example $y^2 = 8x$ satisfies the differential equation (2). This is called *particular solution* a of (2).

In general, the differential equation (4) is formed from (3) by eliminating the arbitrary constant a. Hence (3) is called the *general solution* of the differential equation (4). When a particular value of a is taken it is called a *particular solution.*

WORKED EXAMPLES

Examples 1 Form the differential equations by eliminating the arbitrary constant a. (*i*) $ay^2 = x^3$ (*ii*) $xy = a^2$ (*iii*) $r = a(1-\cos\theta)$

Solution

(i) Now $ay^2 = x^3$

Differentiating w.r.t. 'x' we get $2ay\frac{dy}{dx} = 3x^2$ But $a = \frac{x^3}{y^2}$

Hence substituting the value of a in the above equation we get,

$$2 \cdot \frac{x^3}{y^2} \cdot y \cdot \frac{dy}{dx} = 3x^2$$

Simplifying we get, $2x \cdot \frac{dy}{dx} = 3y$ which is the required differential equation.

(*ii*) Consider $xy = a^2$ Differentiating w.r.t. x, we get

$$x\frac{dy}{dx} + y = 0$$

This is the required differential equation.

(*iii*) Consider $r = a(1-\cos\theta)$. Differentiating w.r.t.θ we get,

$$\therefore \quad \frac{dr}{d\theta} = (0+\sin\theta) = a\sin\theta$$

$$\frac{dr}{d\theta} = \frac{\sin\theta \cdot r}{1-\cos\theta} \quad \text{since } a = \frac{r}{1-\cos\theta} = r \cdot \frac{2\sin\frac{\theta}{2} \cdot \cos\frac{\theta}{2}}{2\sin^2\frac{\theta}{2}}$$

$$\therefore \quad \frac{dr}{d\theta} = r \cdot \cot\frac{\theta}{2}$$

This is the first order differential equation with dependent variable and independent variable θ.

Example 2 Form the differential equations of the circles passing through the origin and having their centres of on (*i*) $x-axis$ (*ii*) $y-axis$ (*iii*) the line $y = x$,

Solution

(i) We know that equation of the circle passing through the origin and having the centre on $x-axis$ is given by

$$x^2 + y^2 + 2gx = 0 \qquad \ldots 1$$

For different values of g we get different circles passing through the origin and having their centres on the $x-axis$. Hence g is the parameter. Differentiating w.r.t. 'x' we get,

$$2x + 2y\frac{dy}{dx} + 2g = 0$$

i.e.,

$$g = -\left(x + y\frac{dy}{dx}\right)$$

Therefore the equation (1) reduces to,

$$x^2 + y^2 - 2x\left(x + y\frac{dy}{dx}\right) = 0$$

i.e., $$\frac{dy}{dx} = \frac{y^2 - x^2}{2xy}$$

Which is the required differential equation.

The equation of a circle passing through the origin and having centre on the y - axis is given by $x^2 + y^2 + 2fy = 0$ ... 2

Here f is the parameter.

Differentiating w.r.t. 'x' we get

$$2x + 2y\frac{dy}{dx} + 2f\frac{dy}{dx} = 0$$

i.e, $$f\frac{dy}{dx} = -\left(x + \frac{dy}{dx}\right)$$

∴ $$f = -\frac{\left(x + \frac{dy}{dx}\right)}{\frac{dy}{dx}}$$

Substituting the value of f in the equation (2) we get,

$$x^2 + y^2 - \frac{2y\left(x + \frac{dy}{dx}\right)}{\frac{dy}{dx}} = 0$$

i.e., $$\left(x^2 + y^2\right)\frac{dy}{dx} - 2y\left(x + \frac{dy}{dx}\right) = 0$$

i.e. $$\left(x^2 + y^2 - 2y\right)\frac{dy}{dx} = 2xy$$

∴ $$\frac{dy}{dx} = \frac{2xy}{x^2 + y^2 - 2y}$$

(*ii*) Now the centre $(-g, -f)$ lies on the line $y = x$ Hence $-f = -g$ or $f = g$

Therefore equation of the circle passing though the origin and having its centre on the line $y = x$ is given by

$$x^2 + y^2 + 2gx + 2gy = 0 \qquad \ldots 3$$

Here we shall eliminate the parameter g. Differentiating w.r.t. x we get.

$$2x + 2y\frac{dy}{dx} + 2g + 2g\frac{dy}{dx} = 0$$

i.e.
$$g\left(1+\frac{dy}{dx}\right) = -\left(x+y\frac{dy}{dx}\right)$$

$\therefore$
$$g = \frac{-\left(x+y\frac{dy}{dx}\right)}{1+\frac{dy}{dx}}$$

Substituting this value of g in the equation (3) we get,

$$x^2 + y^2 - \frac{2(x+y)\left(x+\frac{dy}{dx}\right)}{1+\frac{dy}{dx}} = 0$$

i.e. $$(x^2+y^2)\left(1+\frac{dy}{dx}\right) - 2(x+y)\left(x+\frac{dy}{dx}\right) = 0$$

Simplifying, we, get,

$$\frac{dy}{dx} = \frac{x^2+2xy-y^2}{x^2-2xy-y^2}$$

Example 3 Form the differential equation of the family of confocal ellipses

$$\frac{x^2}{a^2} + \frac{y^2}{a^2+\lambda} = 1$$

By eliminating the parameter λ

Solution Differentiating w.r.t 'x' we get,

$$\frac{2x}{a^2} + \frac{2x}{a^2+\lambda}\cdot\frac{dy}{dx} = 0$$

$$\frac{y}{a^2+\lambda}\cdot\frac{dy}{dx} = \frac{x}{a^2}$$

i.e.
$$\frac{1}{a^2+\lambda} = \frac{x}{a^2+\lambda}\frac{1}{\left(\frac{dy}{dx}\right)}$$

Given equation reduces to,

$$\frac{x^2}{a^2} + y^2\left(-\frac{x}{a^2y}\right)\frac{1}{\frac{dy}{dx}} = 1$$

$$\frac{x^2}{a^2}\cdot\frac{dy}{dx} - \frac{xy}{a^2} = \frac{dy}{dx}$$

$$\left(\frac{x^2}{a^2} - 1\right)\frac{dy}{dx} = \frac{xy}{a^2}$$

$\therefore$
$$\frac{dy}{dx} = \frac{xy}{x^2 - a^2}$$

This is the required differential equation.

Example 4 Form the differential equation of the family of curves $y=ax^2+bx$, where a and b are arbitrary constants.

Solution Since there are two parameters in the given equations, to eliminate them, we have to differentiate twice.
Differentiating the given equation w.r.t. 'x' twice we get,

$$\frac{dy}{dx} = 2ax + b \qquad \text{... i}$$

and
$$\frac{d^2y}{dx^2} = 2a \qquad \text{... ii}$$

Substituting the value of $2a$ from (ii) in (i) we obtain,

$$\frac{dy}{dx} = x\frac{d^2y}{dx^2} + b$$

i.e.
$$b = \frac{dy}{dx} - x\frac{d^2y}{dx^2}$$

Substituting the values of a and b in the given equation we get,

$$y = \frac{x^2}{2}\cdot\frac{d^2y}{dx^2} + \left[\frac{dy}{dx} - x\frac{d^2y}{dx^2}\right]x$$

Simplifying we get,

$$x^2\frac{d^2y}{dx^2} - 2x\frac{dy}{dx} + 2y = 0$$

Which is the required differential equation.

Example 5 Form the differential equation by eliminating the arbitrary constants a and b in the equation $ax^2 + by^2 = 1$

Solution

Given equation can be written as

$$ax^2 + by^2 - 1 = 0 \qquad \text{... 1}$$

Differentiating the equation w.r.t 'x' we get,

$$2ax + 2by\frac{dy}{dx} = 0$$

or

$$ax + by\frac{dy}{dx} = 0 \qquad \ldots 2$$

Differentiating w.r.t. 'x' again , we get,

$$a + b\left\{y\frac{d^2y}{dx^2} + \left(\frac{dy}{dx}\right)^2\right\} = 0 \qquad \ldots 3$$

Eliminating a and b from the equations (1) (2) and (3) we get,

$$\begin{vmatrix} x^2 & y^2 & -1 \\ x & y\frac{dy}{dx} & 0 \\ 1 & y\frac{d^2y}{dx^2} + \left(\frac{dy}{dx}\right)^2 & 0 \end{vmatrix} = 0$$

i.e. $\quad -1\left\{dx\frac{d^2y}{dx^2} + x\left(\frac{dy}{dx}\right)^2 - y\frac{dy}{dx}\right\} = 0$

i.e., $xy\frac{d^2y}{dx^2} + x\left(\frac{dy}{dx}\right)^2 - y\frac{dy}{dx} = 0$, this is the required differential equation.

Example 6 Find the differential equation of the family of curves $xy = Ae^x + Be^{-x} + x^2$ where A and B are arbitrary constants.

Solution

Differentiating the given equation w.r.t. we get,

$$x\frac{dy}{dx} + y = Ae^x - Be^{-x} + 2x \qquad \ldots 1$$

Differentiating w.r.t. 'x' again

$$x\frac{d^2y}{dx^2} + \frac{dy}{dx} + \frac{dy}{dx} = Ae^x + Be^{-x} + 2$$

i.e.,

$$x\frac{d^2y}{dx^2} + 2\frac{dy}{dx} = xy - x^2 + 2$$

using the given equation.

Hence $x\frac{d^2y}{dx^2} + 2\frac{dy}{dx} \; xy = -x^2 + 2$ is the required differential equation.

EXERCISES

1. Form the differential equation by eliminating the arbitrary constant in the following examples.

i. $x^2 = 4ay$ ii. $y = x^2 + ax$

iii. $x^2 + y^2 = ay$ iv. $y^2 = 4a(x+a)$

v. $r = a(1+\cos\theta)$ vi. $r^2 = a^2 \cos 2\theta$

vii. $r = \dfrac{2a}{1+\cos\theta}$

Ans:

i. $\dfrac{dy}{dx} = \dfrac{2y}{x}$ ii. $x\dfrac{dy}{dx} = x^2 + y$

iii. $\dfrac{dy}{dx} = \dfrac{2xy}{x^2 - y^2}$ iv. $y\left(\dfrac{dy}{dx}\right)^2 + 2x\dfrac{dy}{dx} = y$

v. $\dfrac{dr}{d\theta} + r\tan\dfrac{\theta}{2} = 0$ vi. $\dfrac{dr}{d\theta} + r\tan 2\theta = 0$

vii. $\dfrac{dr}{d\theta} = r\tan\dfrac{\theta}{2}$

2. Form the differential equation representing the family of circles passing through the origin and havingthe centers on the line $y = mx$.

$$\left[\textbf{Ans:} \frac{dy}{dx} = \frac{x^2 + 2mxy - y^2}{mx^2 - 2xy - my^2}\right]$$

3. Obtain the differential equation representing the family of rectangular hyperbolas which have coordinates axes as asymptotes.

$$\left[\textbf{Ans:}\, x\frac{dy}{dx} + y = 0\right]$$

4. Form the differential equation representing the family of parabolas having $x - axis$ as the axis and focus at the origin.

$$\left[\textbf{Ans:}\, y\left(\frac{dy}{dx}\right)^2 + 2x\frac{dy}{dx} - y = 0\right]$$

5. Form the differential equation by eliminating the arbitrary constants a from the equation $r^n = a^n cos n\theta$.

$$\left[\textbf{Ans:} \frac{dr}{d\theta} + r\tan n\theta = 0\right]$$

6. Form the differential equation by eliminating the arbitrary constants and in the following families of curves.

   i. $y = a\cos mx + b\sin mx$ ii. $y = ax^2 + bx^3$

   iii. $y = ae^{mx} + be^{\ mx}$ iv. $y = e^x(a\cos x + b\sin x)$

   v. $y = a\sin x + b\cos x + x\sin x$ vi. $y = a\sec x + b\tan x$

Ans:

i. $\dfrac{d^2y}{dx^2} + m^2y = 0$ ii. $\dfrac{d^2y}{dx^2} - m^2y = 0$

iii. $x^2\dfrac{d^2y}{dx^2} - 4x\dfrac{dy}{dx} + 6y = 0$ iv. $\dfrac{d^2y}{dx^2} - 2\dfrac{dy}{dx} + 2y = 0$

v. $\dfrac{d^2y}{dx^2} + y = 2\cos x$ vi. $\dfrac{d^2y}{dx^2} - \tan x\dfrac{dy}{dx} - (\sec^2 x)y = 0$

7. Form the differential equation of all circles with radius r in the xy plane.

Ans: $\left[1+\left(\dfrac{dy}{dx}\right)^2\right]^3 = r^2\left(\dfrac{d^2y}{dx^2}\right)^2$

4.5 Differential Equations of the First Order and of the First Degree

In the preceding section we have considered the method of forming a differential equation by eliminating the arbitrary constant from a given relation. In this section we consider the converse problem that is to find the general solution of a differential equation.

The general form of a differential equation of the first order and of the first degree is of the form

$$\phi\left(x, y\frac{dy}{dx}\right) = 0 \qquad \ldots 1$$

where the degree of $\dfrac{dy}{dx}$ is one.

The general solution of this equation is of the form

$$\psi(x,y,c) = 0 \qquad \ldots 2$$

where c is an arbitrary constant.

There are several methods of solving a differential equation (1). Here we consider the method of solving a differential equation when it is in the *variable separable form*.

4.5.1 Solution of an equation by the method of separation of variables

Suppose a differential equation can be expresses in the form

$$f(x)dx = g(y)dy \qquad \ldots 1$$

Where $f(x)$ is a function of x only and $g(y)$ is a function of y alone. Then we say that the given equation is in the *variable separable form.* To find the solution of such type of equation we integrate w.r.t. the respective variables. Thus the general solution of (1) is given by

$$\int f(x)dx = \int g(y)dy + c$$

Where c is the constant of integration.

If we choose a particular value of c then (2) is called a *particular solution.*

Note:

1. An equation of the form $f_1(x)g_1(y)dx + f_2(x)g_2(y)dy = 0$ can be expresses in the variables separable form by diving throughout the equation by $g_1(y)f_2(x)$.

   That is, $\frac{f_1(x)}{f_2(x)}dx + \frac{g_2(y)}{g_1(y)}dy = 0$. which is in the variable separable form.

2. An equation of the form $f(ax+by+p)dx = g(ax+by+q)dy$ can be reduced to the variable separable form by using the substitution $ax + by = t$

WORKED EXAMPLES

Examples 1 Solve $\sqrt{1-y^2}dx + \sqrt{1-x^2}dy = 0$

Solution Dividing throughout by $\sqrt{1-y^2}\sqrt{1-x^2}$ we get,

$$\frac{dx}{\sqrt{1-x^2}} + \frac{dy}{\sqrt{1-y^2}} = 0$$

On integrating, we get, $\int \frac{dx}{\sqrt{1-x^2}} + \int \frac{dy}{\sqrt{1-y^2}} = c$

i.e., $\sin^{-1} x + \sin^{-1} y = c$ which is the required general equation

Example 2 Solve, $(2y-1)dx-(2x+3)dy=0$.

Solution Dividing by $(2y-1)(2x+3)$, we get,

$$\frac{dx}{2x+3}-\frac{dy}{2y-1} = 0$$

On integrating, we have

$$\int\frac{dx}{2x+3}-\int\frac{dy}{2y-1} = c'$$

$$\frac{1}{2}\log(2x+3)-\frac{1}{2}\log(2y-1) = c'$$

$$\frac{1}{2}\log\frac{2x+3}{2y-1} = c'$$

Since c' is arbitrary we choose $c'=\frac{1}{2}\log c$

$$\therefore \quad \frac{1}{2}\log\frac{2x+3}{2y-1} = \frac{1}{2}\log c$$

$$\frac{2x+3}{2y-1} = c$$

This is the required general solution.

Example 3 Solve $\quad \frac{dy}{dx}+xy = xy^2$

Now $\quad \frac{dy}{dx} = xy^2-xy$

Or $\quad \frac{dy}{dx} = x(y^2-y)$

$$\therefore \quad \frac{dy}{y^2-y} = xdx$$

This is in the variables separable form.

$$\therefore \quad \int\frac{dy}{y(y-1)} = \int xdx+c$$

$$\int\left[\frac{1}{y-1}-\frac{1}{y}\right]dy = \frac{x^2}{2}+c$$

i.e. $$\log(y-1) - \log y = \frac{x^2}{2} + c$$

$$\therefore \quad \log\frac{y-1}{y} = \frac{x^2}{2} + c$$

Which is required general solution.

Example 4 Solve $(y^2 + y) + dx + (x^2 + x)dy = 0$ given that $y = 2$ when $x = 1$

Solution

Dividing throughout by $(y^2 + y)(x^2 + x)$ we, get,

$$\frac{dx}{x^2 + x} + \frac{dy}{y^2 + y} = 0$$

$$\therefore \quad \int\frac{dx}{x(x+1)} + \int\frac{dy}{y(y+1)} = 0$$

i.e. $$\int\left[\frac{1}{x} - \frac{1}{x+1}\right]dx + \int\left[\frac{1}{y} - \frac{1}{y+1}\right]dy = c'$$

$$\log x - \log(x+1) + \log y - \log(y+1) = c'$$

i.e., $$\log\frac{xy}{(x+1)(y+1)} = \log c \quad \text{where } c' = \log c$$

The general solution is $$\frac{xy}{(x+1)(y+1)} = c$$

Given, $x = 1$, $y = 2$ when substituting these values in the above solution, we get,

$$c = \frac{1 \cdot 2}{(1+1)(2+1)} = \frac{1}{3}$$

Hence the particular solution is $\dfrac{xy}{(x+1)(y+1)} = \dfrac{1}{3}$.

Example 5 Solve $\sec^2 x \tan y dx + \sec^2 y \tan x dy = 0$

Solution

Dividing throughout by $\tan x \tan y$ we get, $\dfrac{\sec^2 x}{\tan x}dx + \dfrac{\sec^2 y}{\tan y}dy = 0$

On integrating we get, $\int \frac{\sec^2 x}{\tan x} dx + \int \frac{\sec^2 y}{\tan y} dy = c'$

i.e. $\log \tan x + \log \tan y = c'$
i.e. $\log \tan x + \log \tan y = \log c$ where $c' = \log c$
$\log (\tan x + \tan y) = \log c$
Hence the required general solution is, $\tan x \tan y = c$

Example 6 Solve $y - x\frac{dy}{dx} = a\left(y^2 + \frac{dy}{dx}\right)$

Solution
Now

$$y - x\frac{dy}{dx} = ay^2 + a\frac{dy}{dx}$$

i.e. $$(x+a)\frac{dy}{dx} = y - ay^2$$

$\therefore$ $$\frac{dy}{y-a^2} = \frac{dx}{x+a} + c$$

Hence the variables are seperated
On integrating, we get,

i.e. $$\int \frac{dy}{y-ay^2} = \int \frac{dx}{x+a} + c$$

$$\int \frac{dy}{y(1-ay)} = \int \frac{dy}{x+a} + c'$$

i.e. $$\int \left[\frac{1}{y} + \frac{a}{1-ay}\right] dx = \log(x+a) = c'$$

$$\log y - \log(1-ay) = \log(x+a) + \log c \text{ where } C' = \log c$$

i.e., $$\log \frac{y}{1-ay} = \log c(x+a)$$

Hence the general solution is $\frac{y}{1-ay} = c(x+a)$

Example 7

Solve: (*i*) $\frac{dy}{dx} + 1 = e^{x+y}$ (*ii*) $\frac{dy}{dx} = \sin(x+y)$

Solution

(i) Substitute $x + y = t$,

Differentiating w.r.t. 'x', we get,

$$1+\frac{dy}{dx} = \frac{dt}{dx}$$

Hence the given equation reduces to,

$$\frac{dt}{dx} = e^t \quad \text{or} \quad \frac{dt}{e^t} = dx$$

$$\therefore \quad \int\frac{dt}{e^t} = \int dx + c$$

$$\int e^{-t}dt = \int dx + c$$

i.e., $$-e^{-t} = x + c$$

Hence the general solutions $-e^{-(x+y)} = x + c$

(ii) Substitute $x + y = t$

Differentiating w.r.t., 'x' we get,

$$1+\frac{dy}{dx} = \frac{dt}{dx}$$

$$\frac{dy}{dx} = \frac{dt}{dx} - 1$$

Hence the given equation reduces to,

$$\frac{dt}{dx} - 1 = \sin t$$

i.e., $$\frac{dt}{dx} = 1 + \sin t$$

i.e., $$\frac{dt}{1+\sin t} = dx$$

On integrating w.r.t. 'x', we get

$$\therefore \quad \int\frac{dt}{1+\sin t} = \int dx + c$$

i.e., $$\int\frac{1-\sin t}{1-\sin^2 t}dt = \int dx + c$$

i.e., $$\int\frac{1-\sin t}{\cos^2 t}dt = \int dx + c$$

i.e., $$\int(\sec^2 t - \sec t\tan t)dt = \int dx + c$$

$$\therefore \quad \tan t - \sec t = x + c$$

Hence the general solution is $\tan(x+y) - \sec(x+y) = x + c$

Example 8 Solve $$\frac{dy}{dx} = \frac{2x+3y-4}{4x+6y-1}$$

Solution:

Now

$$\frac{dy}{dx} = \frac{2x+3y-4}{4x+6y-1}$$

$$\frac{dy}{dx} = \frac{2x+3y-4}{2(2x+3y)-1}$$

Substitute, $2x+3y=t$

Therfore, $$2+3\frac{dy}{dx} = \frac{dt}{dx}$$

$$\frac{dy}{dx} = \frac{1}{3}\left(\frac{dt}{dx}-2\right)$$

Given equation reduces to,

$$\frac{1}{3}\left(\frac{dt}{dx}-2\right) = \frac{t-4}{2t-1}$$

$$\frac{1}{3}\frac{dt}{dx} = \frac{t-4}{2t-1}+\frac{2}{3}$$

$$\frac{1}{3}\frac{dt}{dx} = \frac{7t-14}{3(2t-1)}$$

i.e., $$\frac{dt}{dx} = \frac{7(t-2)}{2t-1}$$

i.e., $$\frac{2t-1}{7(t-2)}dt = dx$$

$\therefore$ $$\int\frac{2t-1}{7(t-2)}dt = \int dx+c$$

i.e., $$\frac{1}{7}\int\left[2+\frac{3}{t-2}\right]dt = \int dx+c$$

i.e., $$\frac{1}{7}\int\left[2+\frac{3}{t-2}\right]dt = \int dx+c$$

i.e., $$\frac{1}{7}[2t+3\log(t-2)] = x+c$$

Hence $2(2x+3y)+3\log(2x+3y-2)=7(x+c)$ is the required general solution.

EXERCISES

Solve the following equations

1. $(1+y^2)dx+(1+x^2)dy=0$ [**Ans :** $x+y=c(1-xy)$]

2. $y(1+x)dx+x(1+y)dy=0$ [**Ans :** $\log xy+x+y=c$]

3. $(xy^2+x)dx+(x^2y+y)dy=0$ [**Ans :** $(x^2+1)(y^2+1)=c$]

4. $(2y+1)dx+(3x+1)dy=0$ [**Ans :** $(3x+1)^2(2y+1)^3=c$]

5. $x\sqrt{1+y^2}\,dx+y\sqrt{1+x^2}\,dy=0$ $\left[\textbf{Ans :} \sqrt{1+y^2}+\sqrt{1+x^2}=c\right]$

6. $(e^y+1)\cos x dx+e^y\sin x dy=0$ [**Ans :** $\sin x(e^y+1)=c$]

7. $x\sqrt{y}\,dx+(1+y)\sqrt{1+x}\,dy=0$ [**Ans :** $(x-2)\sqrt{1-x}+(y+3)\sqrt{y}=c$]

8. $\dfrac{dy}{dx}=(y-a)(y-b)$ $\left[\textbf{Ans :} \log\dfrac{y-a}{y-b}=(a-b)x+c\right]$

9. $\dfrac{dy}{dx}+xy=xy^3$ $\left[\textbf{Ans :} \log\dfrac{y^2-1}{y^2}=x^2+c\right]$

10. $\cos y\log(\sec x+\tan x)dx=\cos x\log(\sec y+\tan y)dy$
[**Ans :** $[\log(\sec x+\tan x)]^2-[\log(\sec y+\tan y)]^2=c$]

11. $\left(y^2+y+1\right)dx-(x^2+x+1)dy=0$
$\left[\textbf{Ans :} \tan^{-1}\dfrac{2x+1}{\sqrt{3}}-\tan^{-1}\dfrac{2y+1}{\sqrt{3}}=c\right]$

12. $\dfrac{dy}{dx}=\dfrac{x(2\log+1)}{\sin y+y\cos y}$ [**Ans :** $y\sin=x^2\log x+c$]

13. $y-x\dfrac{dy}{dx}=a\left(x^2\dfrac{dy}{dx}+1\right)$ [**Ans :** $x=c(y-a)(1+x)$]

14. $(1-x^2)(1-y)dx=xy(1+y)dy$
$\left[\textbf{Ans :} \log x-\dfrac{x^2}{2}=-\dfrac{y^2}{2}-2y-2\log(1-y)+c\right]$

15. $a\left(x\dfrac{dy}{dx}+2y\right)=xy\dfrac{dy}{dx}$ $\left[\textbf{Ans :} \log x^2y=\dfrac{y+c}{a}\right]$

16. $\dfrac{dy}{dx}=(x+y)^2$ [**Ans :** $\tan^{-1}(x+y)=x+c$]

17. $\frac{dy}{dx} = \frac{1}{\cos(x+y)}$ [**Ans :** $x + y + \cot(x+y) - \operatorname{cosec}(x+y) = x + c$]

18. $\frac{dy}{dx} = \sin(x+y) + \cos(x+y)$ $\left[\textbf{Ans :} 1 + \tan\frac{x+y}{2} = e^x c\right]$

19. $(x+y+1)\frac{dy}{dx} = 1$ [**Ans :** $x + y + 2 = ce^y$]

20. $\frac{dy}{dx} = 1 + 6xe^{x-y}$ [**Ans :** $e^{-(x-y)} = 3x^2 + c$]

21. $e^y\left(\frac{dy}{dx} - 1\right) = e^x$ $\left[\textbf{Ans :} e^{-(x-y)} = x + c\right]$

22. $\frac{dy}{dx} = \frac{x - 2y + 1}{2x - 4y + 3}$ $\left[\textbf{Ans :} \frac{(2x-4y+3)^2}{4} = x + c\right]$

23. $\frac{dy}{dx} = (x+y+1)^2$ [**Ans :** $\tan^{-1}(x+y+1) = x + c$]

24. $\frac{y}{x}\cdot\frac{dy}{dx} + \frac{x^2 + y^2 - 1}{2(x^2+y^2)+1} = 0$ Hint : Substitute $x^2 + y^2 = t$

[**Ans :** $x^2 + 2y^2 = 3\log(x^2 + y^2 + 2) + c$]

4.5.2 Homogeneous equations

Suppose the differential equation to be solved is of the form

$$\frac{dy}{dx} = f(x, y) \quad \ldots 1$$

where $f(x, y)$ is a function of x *and* y which can be re written as a function of y/x or as a function of x/y. (such function are called *homogenous functions* of x *and* y of degree zero). Then (1) is called a *homogenous differential equation.*

If $f(x, y)$ is rewritten as a function of y/x, then we put $y = vx$, where $v = v(x)$ so that

$$\frac{dy}{dx} = v + x\frac{dv}{dx} \quad \ldots 2$$

Substituting for y and $\frac{dy}{dx}$ into (1), we get an equation involving v and x, which may be solved by separating the variables.

On the other hand, if $f(x, y)$ can be written as a function of x/y, then we put $x = vy$, where $v = v(y)$, so that $\frac{dy}{dx} = v + y\frac{dv}{dy}$ ... 3

Then (1) becomes an equation an equation involving v *and* y, which may be solved by separating the variables.

WORKED EXAMPLES

Examples 1 Solve $\dfrac{dy}{dx} = \dfrac{y}{x} + \sin\dfrac{y}{x}$

Solution Since the right hand side is a function of (y/x) the given equation is homogeneous. Setting $y = vx$,

so that $\dfrac{dy}{dx} = v + x\dfrac{dv}{dx}$ the equation becomes,

$v + x\dfrac{dv}{dx} = v + \sin v$ or $\operatorname{cosec} v\, dv = \dfrac{dx}{x}$

Integrating this, we get,

$\log(\operatorname{cosec} v - \cot v) = \log x + \text{constant}$

$$\text{or} \quad \log\left[\frac{\operatorname{cosec} v - \cot v}{x}\right] = \text{constant}$$

$$\text{or} \quad \frac{1}{x}(\operatorname{cosec} v - \cot v) = c, a \text{constant}$$

Substituting for v, this becomes

$$\operatorname{cosec}(y/x) - \cot(y/x) = cx$$

This is the general solutions of the given equation.

Example 2 Solve $x\dfrac{dy}{dx} = y(\log y - \log x + 1)$

Solution The given equation may be written as

$$\frac{dy}{dx} = \frac{y}{x}\left(\log\frac{y}{x} + 1\right)$$

which is in homogeneous form. Setting $y = vx$, this equation becomes

$v + x\dfrac{dv}{dx} = v(\log v + 1)$, or $\quad \dfrac{(1/v)dv}{\log v} = \dfrac{dx}{x}$

Integrating this we get,

$$\log(\log v) = \log x + \text{constant}$$

$$\text{or} \quad \log\left(\frac{\log v}{x}\right) = \text{constant}$$

Substituting for *v*, this becomes

$$\log\left(\frac{y}{x}\right) = cx$$

This is the general solutions of the given equation.

Example 3 Solve $(x^2 - xy + y^2)dx - xydy = 0$

Solution The given equation may be written as

$$\frac{dy}{dx} = \frac{x^2 - xy + y^2}{xy}$$

$$= \frac{x^2[1-(y/x)+(y^2/x^2)]}{xy}$$

$$= \frac{1-(y/x)+(y/x)^2]}{(y/x)}$$

which is in homogeneous form. Setting $y = vx$, this equation becomes

$$v + x\frac{dv}{dx} = \frac{1-v+v^2}{v}$$

or $$x\frac{dv}{dx} = \frac{1-v+v^2}{v} - v = \frac{1-v}{v}$$

or $$\frac{dx}{x} = \frac{v}{1-v}dv = \frac{(v-1)+1}{1-v}dv = -\left[1+\frac{1}{v-1}\right]dv$$

Integrating this, we get

$$\int\frac{1}{x}dx = -\int\left[1+\frac{1}{v-1}\right]dv$$

$$\log x = -[v + \log(v-1)] + c$$

$$\log x(v-1) = c - v$$

Recalling that $v = \frac{y}{x}$, this becomes

$$\log(y - x) = c - \left(\frac{y}{x}\right)$$

This is the general solution of the given equation.

Example 4 Solve $x^2ydx - (x^3 + y^3)dy = 0$

Solution The given equation may be written as

$$\frac{dy}{dx} = \frac{x^2y}{x^3+y^3} = \frac{y/x}{1+(y/x)^3}$$

which is in homogeneous form. Setting $y = vx$ this equation becomes

$$v + x\frac{dv}{dx} = \frac{v}{1+v^3}$$

or $$x\frac{dv}{dx} = \frac{v}{1+v^3} - v = \frac{-v^4}{1+v^3}$$

or $$= \frac{1+v^3}{v^4}dv = -\frac{dx}{x}$$

Integrating this, we get

$$\int \frac{1}{v^4}dv + \int \frac{1}{v}dv = -\int \frac{1}{x}dx$$

$$-\frac{1}{3}\cdot\frac{1}{v^3} + \log v = -\log x + c$$

$$3\log vx - \frac{1}{v^3} = c$$

Recalling that $v = \frac{y}{x}$, this becomes

$$3\log y - \frac{x^3}{y^3} = c$$

This is the general solution of the given equation.

Example 5 Solve $(y^2 + 2xy)dx + (2x^2 + 3xy)dy = 0$

Solution The given equation may be written as

$$\frac{dy}{dx} = -\frac{y^2 + 2xy}{2x^2 + 3xy}$$

$$= -\frac{\left(\frac{y}{x}\right)^2 + 2\left(\frac{y}{x}\right)}{2 + 3(y/x)}$$

which is in homogeneous form. Setting $y = vx$ this equation becomes

$$v + x\frac{dv}{dx} = -\frac{v^2 + 2v}{2 + 3v}$$

or $$x\frac{dv}{dx} = -\frac{v^2 + 2v}{2 + 3v} - v$$

$$= -4\frac{v^2 + v}{2 + 3v}$$

Separating the variables, this becomes

$$-4\frac{dx}{x} = \frac{2+3v}{v^2+v}dv$$

$$\frac{-4}{x}dx = \left[\frac{2}{v}+\frac{1}{v+1}\right]dv \text{, by partial fractions}$$

Integration this, we get

$$-4(\log x) + c = 2\log v + \log(v+1) \text{ or}$$

$$\log v^2 (v+1)x^4 = c$$

Recalling that $xv = y$, this becomes

$$\log xy^2(x+y) = c$$

This is the general solution of the given equation.

Example 6 $\left[x\tan\left(\frac{y}{x}\right) - y\sec^2\left(\frac{y}{x}\right)\right]dx + x\sec^2\left(\frac{y}{x}\right)dy = 0$

Solution The given equation may be written as

$$\frac{dy}{dx} = \frac{\left(\frac{y}{x}\right)\sec^2\left(\frac{y}{x}\right) - \tan\left(\frac{y}{x}\right)}{\sec^2\left(\frac{y}{x}\right)}$$

which is in homogeneous form. Setting $y = vx$, this equation becomes

$$v + x\frac{dv}{dx} = \frac{v\sec^2 v - \tan v}{\sec^2 v}$$

Or $$x\frac{dv}{dx} = \frac{v\sec^2 v - \tan v}{\sec^2 v} - v = -\frac{\tan v}{\sec^2 v}$$

Separating the variables, this becomes

$$= \frac{\sec^2 v}{\tan v}dv = -\frac{dx}{x}$$

Integrating this, we get

$$= \int\frac{\sec^2 v}{\tan v}dv = -\int\frac{dx}{x} + c$$

$$\log(\tan v) = -\log x + c$$

or $$\log(x\tan v) = c \quad \text{or} \quad x\tan v = c$$

Recalling that $xv = y$, this becomes

$$x\tan y/x = c$$

This is the general solution of the given equation.

Example 7 Solve $xdy - y\,dx = \sqrt{x^2 + y^2}\,dx$

The given equation may be written as

$$x\frac{dy}{dx} = y + \sqrt{x^2 + y^2}$$

$$\frac{dy}{dx} = \frac{y}{x} + \frac{\sqrt{x^2 + y^2}}{x}$$

$$\frac{dy}{dx} = \frac{y}{x} + \sqrt{1 + \frac{y^2}{x^2}}$$

which is in homogeneous form. Setting $y = vx$ this equation becomes

$$v + x\frac{dv}{dx} = v + \sqrt{1 + v^2}$$

or

$$x\frac{dv}{dx} = v + \sqrt{1 + v^2} - v$$

$$= \sqrt{1 + v^2}$$

Separating the variables, this becomes

$$\frac{dv}{\sqrt{1 + v^2}}dv = \frac{dx}{x}$$

Integrating this, we get

$$\int \frac{dv}{\sqrt{1 + v^2}} = \int \frac{dx}{x} + c$$

$$\log\left[v + \sqrt{1 + v^2}\right] = \log x + c$$

$$\frac{v + \sqrt{1 + v^2}}{x} = c$$

$$v + \sqrt{1 + v^2} = cx$$

Recalling that $xv = y$, this becomes

$$y + \sqrt{x^2 + y^2} = cx^2$$

This is the general solution of the given equation.

Example 8 Solve $(x + y)dy + (x - y)dx = 0$

Solution The given equation may be written as

$$\frac{dy}{dx} = \frac{y - x}{y + x}$$

$$\frac{dy}{dx} = \frac{\left(\frac{y}{x}\right)-1}{\left(\frac{y}{x}\right)+1}$$

which is in homogeneous form. Setting $y = vx$ this equation becomes

$$v + x\frac{dv}{dx} = \frac{v-1}{v+1}$$

or

$$x\frac{dv}{dx} = \frac{v-1}{v+1} - v = -\frac{v^2+1}{v+1}$$

Separating the variables, this becomes

$$\frac{v+1}{v^2+1}dv = -\frac{dx}{x}$$

Integrating this, we get

$$\int \frac{v}{v^2+1}dv + \int \frac{1}{v^2+1}dv = -\int \frac{dx}{x} + c$$

$$\frac{1}{2}\log(v^2+1) + \tan^{-1} v = -\log x + c$$

$$\log x\sqrt{v^2+1} + \tan^{-1} v = c$$

Recalling that $v = \frac{y}{x}$, this becomes

$$\log\sqrt{y^2+x^2} + \tan^{-1}(y/x) = c$$

This is the general solution of the given equation.

Example 9 Solve $\frac{dy}{dx} = \frac{x+2y}{2x+y}$

The given equation may be written as

$$\frac{dy}{dx} = \frac{1+2\left(\frac{y}{x}\right)}{2+\left(\frac{y}{x}\right)}$$

which is in homogeneous form. Setting $y = vx$, this equation becomes

$$v + x\frac{dv}{dx} = \frac{1+2v}{2+v}$$

or

$$x\frac{dv}{dx} = \frac{1+2v}{2+v} - v = \frac{1-v^2}{2+v}$$

Separating the variables, this becomes

$$\frac{2+v}{1-v^2}dv = \frac{dx}{x}$$

$$\frac{dx}{x} = \frac{2+v}{1-v^2}dv = -\frac{1}{2}\left[\frac{3}{v-1}-\frac{1}{v+1}\right]$$

Integrating this, we get

$$-\frac{1}{2}\left[\int\frac{3}{v-1}dv-\int\frac{1}{v+1}dv\right] = \int\frac{dx}{x}+c$$

$$-\frac{1}{2}[3\log(v-1)-\log(v+1)] = \log x + c$$

$$-\left[\log(v-1)^{3/2}-\log(v+1)^{1/2}\right] = \log x + c$$

$$-\left[\log\frac{(v-1)^{3/2}}{(v+1)^{1/2}}\right] = \log x + c$$

$$\log\left\{x\frac{(v-1)^{3/2}}{(v+1)^{1/2}}\right\} = c = \log c$$

$$x\frac{(v-1)^{3/2}}{(v+1)^{1/2}} = c$$

Recalling that $v=\frac{y}{x}$, this becomes

$$\frac{(y-x)^{3/2}}{(y+x)^{1/2}} = c \text{ or } (y-x)^3 = c^2 (y+x)$$

This is the general solution of the given equation.

Example 10 Solve $y^2+(y^2-xy)\frac{dy}{dx}=0$

Solution

The given equation may be written as

$$\frac{dx}{dy} = \frac{xy-x^2}{y^2} = \frac{x}{y}-\frac{x^2}{y^2} \qquad \ldots \text{i}$$

The right-hand side of this equation is a homogeneous fuction.
Therefore, which is in homogeneous form.

Setting $x = vy$ this equation becomes

$$\frac{dx}{dy} = y\frac{dv}{dy} + v$$

Then the equation (i) becomes

$$y\frac{dv}{dy} + v = v - v^2$$

or

$$y\frac{dv}{dy} = -v^2$$

Separating the variables, this becomes

$$-\frac{1}{v^2}dv = \frac{dy}{y}$$

Integrating this, we get

$$-\int\frac{1}{v^2}dv = \int\frac{dy}{y} + c$$

$$\frac{1}{v} = \log y + c$$

Recalling that $v = \frac{x}{y}$, this becomes

$$\frac{y}{x} = \log y + c$$

This is the general solution of the given equation.

Example 11 Solve $(1 + e^{x/y})dx + e^{x/y}(1 - x/y)dy = 0$

Solution The given equation may be written as

$$\frac{dx}{dy} = -\frac{e^{x/y}(1 - x/y)}{1 + e^{x/y}}$$

which is in homogeneous form. Setting $x = vy$ this equation becomes

$$\frac{dx}{dy} = y\frac{dv}{dx} + v$$

and the equation becomes

$$v + y\frac{dv}{dx} = -\frac{e^v(1 - v)}{1 + e^v}$$

or

$$y\frac{dv}{dy} = -\frac{e^v(1 - v)}{1 + e^v} - v = -\frac{v + e^v}{1 + e^v}$$

Separating the variables, this becomes

$$\frac{1+e^v}{v+e^v}dv = -\frac{dy}{y}$$

Integrating this, we get

$$\int \frac{1+e^v}{v+e^v}dv = -\int \frac{dy}{y} + c$$

$$\log(v + e^v) = -\log y + c$$

$$\log y(v + e^v) = c \quad \text{or} \quad y(v + e^v) = c$$

Recalling that $yv = x$, this becomes

$$x + ye^{x/y} = c$$

This is the general solution of the given equation.

Example 12 Solve $\left(x + y\cot\frac{x}{y}\right)dy - ydx = 0$

Solution

The given equation may be written as

$$\frac{dx}{dy} = \frac{x}{y} + \cot\frac{x}{y}$$

which is in homogeneous form. Setting $x = vy$ this equation becomes

$$\frac{dx}{dy} = v + y\frac{dv}{dy}$$

and the equation becomes

$$v + y\frac{dv}{dx} = v + \cot v$$

or $$y\frac{dv}{dx} = v + \cot v - v = \cot v$$

Separating the variables, this becomes

$$\frac{1}{\cot v}dv = \frac{dy}{y} \quad \text{or} \quad \tan v\, dv = \frac{dy}{y}$$

Integrating this, we get

$$\int \tan v\, dv = \int \frac{dy}{y} + c$$

$$\log\sec v = \log y + c$$

$$\log\frac{\sec v}{y} = c \quad \text{or} \quad \frac{\sec v}{y} = c$$

Recalling that $yv = x$, this becomes

or $$\sec\frac{x}{y} = cy$$

This is the general solution of the given equation.

EXERCISES

Solve the following equations

1. $\frac{dy}{dx} = \frac{x^2 + y^2}{2x^2}$ **[Ans :** $2x = (x - y)(\log x + c)$**]**

2. $\frac{dy}{dx} = \frac{x^2 - y^2}{2xy}$ **[Ans :** $x \,.\, (x^2 - 3y^2) = c$**]**

3. $\frac{dy}{dx} = \frac{x^2 + 3y^2}{2xy}$ **[Ans :** $x^2 + y^2 = cx^3$**]**

4. $\frac{dy}{dx} = \frac{y^2 - 2xy^2}{x^2 - 2xy}$ **[Ans :** $xy(x - y) = c$**]**

5. $\frac{dy}{dx} = \frac{x^3 + 2y^3}{2xy^2}$ **[Ans :** $2y^3 = 3x^3(\log x + c)$**]**

6. $\frac{x}{y}\frac{dy}{dx} = \frac{x\cos(y/x) + y\sin(y/x)}{y\sin(y/x) - x\cos(y/x)}$ $\left[\textbf{Ans}: xy\cos\left(\frac{y}{x}\right) = c\right]$

7. $x(x - y)dy + y^2 dx = 0$ $\left[\textbf{Ans}: y = e^{\frac{y}{x}} = c\right]$

8. $x^2 dx + y(x + y)dx = 0$ **[Ans :** $y + 2x = cyx^2$**]**

9. $(xy + y^2)dx + (x^2 - xy)dx = 0$ **[Ans :** $xy + ce^{(y/x)}$**]**

10. $(x^3 + y^3)dx = (x^2 y + xy^2)dx = 0$ **[Ans :** $(y/x) + \log(y - x) = c$**]**

11. $(2xy^2 - x^3)dy = 3xy(ydx - xdy)$ **[Ans :** $(x^2 + y^2)^4 = c(y^2 - x^2)^2$**]**

12. $x^3 dx - y^3 dy = 3xy(ydx - xdy)$ **[Ans :** $(x^2 + y^2)^4 = c(y^2 - x^2)^2$**]**

13. $\frac{dy}{dx} = y + x\cos^2\left(\frac{y}{x}\right)$ **[Ans :** $\log x = \tan(y/x) + c$**]**

14. $\frac{dy}{dx} = \frac{y}{x} + \tan\frac{y}{x}$ **[Ans :** $\sin(y/x) = cx$**]**

15. $x\frac{dy}{dx} + \frac{y^2}{x} = y$ **[Ans :** $(-x/y) + \log(y - x) + c$**]**

16. $y^2 + x^2\frac{dy}{dx} = xy\frac{dy}{dx}$ **[Ans :** $y^2 = 2x[y + x\log(cx)]$**]**

17. $y - x\frac{dy}{dx} = x + y + \frac{dy}{dx}$ $\left[\textbf{Ans}: \log\sqrt{x^2 + y^2} + \tan^{-1}(y/x) = c\right]$

18. $\dfrac{dy}{dx}=\dfrac{y+\sqrt{y^2+x^2}}{x}$ $\left[\textbf{Ans : } y+\sqrt{x^2+y^2}=cx^2\right]$

19. $\dfrac{dy}{dx}=1+\dfrac{y}{x}+\dfrac{y^2}{x^2}$ [**Ans :** $y=x\tan\log(cx)$]

20. $ye^{x/y}dx=(xe^{x/y}+y^2)dy$ $\left[\textbf{Ans : } e^{(y/x)}=y+c\right]$

21. $xdx+\sin^2\left(\dfrac{y}{x}\right)(ydx-xdy)=0$ $\left[\textbf{Ans : } \log x=\dfrac{1}{2}\left[\dfrac{y}{x}-\dfrac{1}{2}\sin\left(\dfrac{2y}{x}\right)\right]+c\right]$

22. $xy\log(x/y)dx+[y^2-x^2\log(x/y)]dy=0$

$$\left[\textbf{Ans : } \log y-\left(\frac{x^2}{4y^2}\right)\left[2\log\left(\frac{y}{x}\right)+1\right]=c\right]$$

4.5.3 Equations reducible to homogeneous form

Suppose the equation to be solved is of the form

$$\frac{dy}{dx} = \frac{ax+by+c}{Ax+By+C} \qquad \dots 1$$

Where *a, b, c* are *A, B, C,* constants. We consider two cases.

Case (i): Suppose the constants *a, b A, B* are such that $ax+by=m(Ax+By)$ where *m* is a constant. Then the equation (1) becomes

$$\frac{dy}{dx} = \frac{m(Ax+By)+c}{Ax+By+C} \qquad \dots 2$$

To solve this equation, we put

$$Ax+By = t \qquad \dots 3$$

So that $$A+B\frac{dy}{dx} = \frac{dt}{dx}$$

Then the equation (2) can rewritten as

$$\frac{dt}{dx} = A+B\left(\frac{mt+c}{t+C}\right) \qquad \dots 4$$

This equation can be solved by separating the variables.

Similar procedure is adopted when $Ax+By=n(ax+by)$ for some constant *n*.

Case (ii): Suppose $ax+by\neq m(Ax+By)$ for any constant *m*.

In this case, we put

$$x = X+h, \quad y=Y+k \qquad \dots 5$$

Where *h* and *k* are constants. Then

$$\frac{dy}{dx} = \frac{dy}{dY}\frac{dY}{dX}\cdot\frac{dX}{dx}=1\cdot\frac{dY}{dX}=\frac{dY}{dX} \qquad \dots 6$$

and equation (1) becomes

$$\frac{dY}{dX} = \frac{aX + bY + (ah + bk + c)}{AX + BY + (Ah + Bh + C)} \qquad \ldots 7$$

We choose the constants h and k such that

$ah + bk + c = 0, \quad Ah + Bk + C = 0$ $\qquad \ldots 8$

Then (7) reduces to

$$\frac{dY}{dX} = \frac{aX + bY}{AX + BY} \qquad \ldots 9$$

This is a homogeneous equation which can be solved by taking $Y = VX$ where $V = V(X)$ or by taking $X = VY$, where $V = V(Y)$

WORKED EXAMPLES

Example 1 Solve $(4x - 6y - 1)dx - (2x - 3y + 2)dy = 0$

Solution The given equation may be written as

$$\frac{dy}{dx} = \frac{2(2x+3y)-1}{2x-3y+2} \qquad \ldots 1$$

which is in the form of (2). To solve the equation, we take

$$2x - 3y = t$$

So that

$$2 - 3\frac{dy}{dx} = \frac{dt}{dx}$$

Then equation (1) gives

$$\frac{dt}{dx} = 2 - 3\left(\frac{2t-1}{t+2}\right) = -\frac{4t-7}{t+2}$$

$$dx = -\frac{t+2}{4t-7}dt = -\frac{1}{4}\left[1 + \frac{15}{4t-7}\right]dt$$

on intergrating, we get

$$\int dx = -\frac{1}{4}\left[\int 1dt \int \frac{15}{4t-7}dt\right] + c$$

$$x = -\frac{1}{4}\left[t + \frac{15}{4}\log\left(t - \frac{7}{4}\right)\right] + c$$

Substituting for t from (2), this becomes

$$x = -\frac{1}{4}\left[2x - 3y + \frac{15}{4}\log\left(2x - 3y - \frac{7}{4}\right)\right] + c$$

Example 2 Solve $(x + 2y)(dx - dy) = dx + dy$

Solution The given equation may be written as
$(x + 2y - 1)\, dx = (x + 2y + 1)dy$

$$\frac{dy}{dx} = \frac{x+2y-1}{x+2y+1} \qquad \ldots 1$$

which is in the form of (2). To solve the equation, we take

$$x + 2y = t \qquad \ldots 2$$

So that $1+2\dfrac{dy}{dx}=\dfrac{dt}{dx}$, Then equation (1) gives

$$\frac{dt}{dx} = 1+2\left(\frac{t-1}{t+1}\right)=\frac{3t-1}{t+1}$$

$$dx = \frac{t+1}{3t-1}dt=\frac{1}{3}\left[1+\frac{4}{3t-1}\right]dt$$

on intergrating, we get

$$\int dx = \frac{1}{3}\left[\int 1dt \int \frac{4}{3t-1}dt\right]+c$$

$$x = \frac{1}{3}\left[t+\frac{4}{3}\log\left(t-\frac{1}{3}\right)\right]+c$$

Substituting for t from (2), this becomes

$$x = \frac{1}{3}\left[x+2y+\frac{4}{3}\log\left(x+2y-\frac{1}{3}\right)\right]+c$$

Example 3 Solve $\dfrac{dy}{dx}=\dfrac{y-x+1}{y+x+5}$

Soltution To solve the given equation, we put
$x = X + h, \quad y = Y + k$ ... 1
Then the equation becomes

$$\frac{dY}{dX} = \frac{Y-X+(k-h+1)}{Y+X+(k+h+5)} \qquad \ldots 2$$

We choose the constants h and k such that
$k - h + 1 = 0, \quad k + h + 5 = 0$
These equations yield $k = -3, \quad h = -2$ ... 3
which is in the form of (2) reduces to

$$\frac{dY}{dX} = \frac{Y-X}{Y+X} \qquad \ldots 4$$

which is Homogeneous equation . we take

$$Y = VX \qquad \dots 5$$

So that equation (4) becomes

$$V + X\frac{dV}{dX} = \frac{V-1}{V+1}$$

$$X\frac{dV}{dX} = \frac{V-1}{V+1} - V = -\frac{V^2+1}{V+1}$$

$$\frac{dX}{X} = -\frac{V+1}{V^2+1}dV$$

$$= -\frac{1}{2}\frac{2V}{V^2+1}dV - \frac{1}{V^2+1}dV$$

on intergrating, we get

$$\int\frac{dX}{X} = -\frac{1}{2}\int\frac{2V}{V^2+1}dV - \int\frac{1}{V^2+1}dV + c$$

$$\log X = -\frac{1}{2}\log(V^2+1) - \tan^{-1}V + c$$

$$\log(V^2+1)^{1/2}X + \tan^{-1}V = c$$

Equation (5) becomes

$$\log(X^2+Y^2)^{1/2} - \tan^{-1}\left(\frac{Y}{X}\right) = c$$

Substituting for X and Y from (1) and using (3) this becomes

$$\log\{(x+2)^2 + (y+3)^2\}^{1/2} - \tan^{-1}\left(\frac{y+3}{x+2}\right) = c$$

This is the general solution of the given equation.

Example 4 Solve $\dfrac{dy}{dx} = \dfrac{x+y-1}{x-y+1}$

To solve the given equation, we put

$x = X + h, \quad y = Y + k$

Then the equation becomes

$$\frac{dY}{dX} = \frac{X+Y+(h+k-1)}{X-Y+(h-k+1)} \qquad \dots 1$$

We choose the constants h and k such that

$h + k - 1 = 0, \quad h - k + 1 = 0$ ... 2

Solving these equations, we get

$h = 0, \quad k = 1$

which is in the form of (1) reduces to

$$\frac{dY}{dX} = \frac{X+Y}{Y-X} = \frac{1+\frac{Y}{X}}{1-\frac{Y}{X}}$$

Which is Homogeneous equation. we take $Y = VX$

This equation becomes

$$V + X\frac{dV}{dX} = \frac{1+V}{1-V}$$

$$X\frac{dV}{dX} = \frac{1+V}{1-V} - V = \frac{1+V^2}{1-V}$$

$$= \frac{1-V}{1+V^2}dV = \frac{dX}{X}$$

On intergrating we get

$$\int\frac{dX}{X} = \int\frac{1}{1+V^2}dV - \frac{1}{2}\int\frac{2V}{1+V^2}dV + c$$

$$\log X = \tan^{-1}V - \frac{1}{2}\log(1+V^2) + c$$

$$\tan^{-1}V - \log(1+V^2)^{1/2}X = c$$

Substituting for V and then X and Y, this becomes

$$\tan^{-1}\left(\frac{Y}{X}\right) - \log\left(X^2+Y^2\right)^{1/2} = c$$

$$\tan^{-1}\left(\frac{y-1}{x}\right) - \log\left[x^2+(y-1)^2\right]^{1/2} = c$$

This is the general solution of the given equation.

Example 5 Solve $(3y - 7x + 7)dx + (7y - 3x + 3)dy = 0$

Solution The given equation is

$$\frac{dy}{dx} = -\frac{3y-7x+7}{7y-3x+3} = \frac{7x-3y-7}{7y-3x+3} \quad \ldots 1$$

To solve the given equation, we put

$x = X + h, \quad y = Y + k$ ... 2

the equation (1) becomes

$$\frac{dY}{dX} = \frac{7X-3Y+(7h-3k-7)}{7Y-3X+(7k-3h+3)} \quad \ldots 3$$

We choose the constant h and k such that

$7h - 3k - 7 = 0, \quad 7k - 3h + 3 = 0$ ... 2

Solving these equations , we get

$h = 1, \quad k = 0$

which is in the form of (3) reduces to

$$\frac{dY}{dX} = \frac{7X - 3Y}{7Y - 3X}$$

Which is Homogeneous equation. we take $Y = VX$, this reduses to This equation becomes

$$V + X\frac{dV}{dX} = \frac{7 - 3V}{7V - 3}$$

$$X\frac{dV}{dX} = \frac{7 - 3V}{7V - 3} - V = \frac{7(1 - V^2)}{7V - 3}$$

$$\frac{dX}{X} = \frac{7V - 3}{7(1 - V^2)}dV$$

$$= -\frac{1}{7}\left[\frac{5}{V + 1} + \frac{2}{V - 1}\right]$$

On intergrating, we get

$$\int\frac{dX}{X} = -\frac{1}{7}\left[\int\frac{5}{V + 1}dV + \int\frac{2}{V - 1}dV\right] + c$$

$$\log X = -\frac{1}{7}[5\log(V + 1) + 2\log(V - 1)] + c$$

$$7\log X = [5\log(V + 1) + 2\log(V - 1)] + 7c$$

$$X^7(V + 1)^5(V - 1)^2 = c$$

Substituting for $V = \frac{Y}{X}$, this becomes,

$$(Y + X)^5(Y - X)^2 = c$$

Using (2) and (3), this reduces to

$$(y + x - 1)^5(y - x + 1)^2 = c$$

This is the general solution of the given equation.

Example 6 Solve $\frac{dy}{dx} = \frac{x + 2y - 3}{2x + y - 3}$

Solution To solve the given equation, we put

$x = X + h, \quad y = Y + k$

Then the equation becomes

$$\frac{dY}{dX} = \frac{X+2Y+(h+2k-3)}{2X+Y+(2h+k-3)} \qquad \ldots 1$$

We choose the constants h and k such that

$h+2k-3=0, \quad 2h+k-3=0$

Solving these equations, we get

$h=1, \; k=1 \qquad \ldots 2$

which is in the form of reduces to

$$\frac{dY}{dX} = \frac{X+2Y}{2X+Y} = \frac{1+2\left(\frac{Y}{X}\right)}{2+\left(\frac{Y}{X}\right)}$$

Which is Homogeneous equation. we take $Y = VX$

This equation becomes

$$V + X\frac{dV}{dX} = \frac{1+2V}{2+V}$$

$$X\frac{dV}{dX} = \frac{1+2V}{2+V} - V = \frac{1-V^2}{2+V}$$

$$\frac{dX}{X} = -\frac{2+V}{V^2-1}dV = -\frac{1}{2}\left[\frac{3}{V-1} - \frac{1}{V+1}\right]dV$$

On integrating, we get

$$\int\frac{dX}{X} = -\frac{1}{2}\left[\int\frac{3}{V-1}dV - \int\frac{1}{V+1}dV\right] + c$$

$$\log X = -\frac{1}{2}[3\log(V-1) - \log(V+1) + c]$$

$$\log X\frac{(V-1)^{3/2}}{(V+1)^{1/2}} = c$$

$$X\frac{(V-1)^{3/2}}{(V+1)^{1/2}} = c$$

Substituting for V and then X and Y, this become

$$(y-x)^3 = c^2\,(y+x-2)$$

This is the general solution of the given equation.

EXERCISES

1. $\dfrac{dy}{dx} = \dfrac{x-y+3}{2x-2y+5}$ [**Ans :** $\log(x-y+2) = 2y-x+c$]

2. $\frac{dy}{dx} = \frac{x+y+1}{2x+2y+3}$ [**Ans :** $3(2y-x)+\log(3x+3y+4)$]
3. $\frac{dy}{dx} = \frac{y-x+1}{y-x+5}$ [**Ans :** $(y-x)^2+10y-2x=c$]
4. $\frac{dy}{dx} = \frac{x-y}{x-8y+7}$ [**Ans :** $x^2-2xy+8y^2+14y=c$]
5. $\frac{dy}{dx} = \frac{y+x-2}{y-x-4}$ [**Ans :** $x^2+2xy-y^2-4x+8y-14=c$]
6. $\frac{dy}{dx} = \frac{2x+5y+1}{5x+2y-1}$ [**Ans :** $(x+y)^7 = c\,(x-y-2/3)^3$]
7. $\frac{dy}{dx} = \frac{2(y-x)}{3x-y-1}$ [**Ans :** $(x+y-1)^4 = c\,(4x-2y-1)$]
8. $\frac{dy}{dx} = \frac{x-y+1}{x+2y-3}$ $\left[\textbf{Ans:} \left(x-\frac{1}{3}\right)^2 - 2\left(x-\frac{1}{3}\right)\left(y-\frac{4}{3}\right) - 2\left(y-\frac{4}{3}\right)^2 = c\right]$
9. $(x+y)(dx-dy) = dx+dy$ [**Ans :** $x-y=\log(x+y)+c$
10. $(2x+y+1)dx+(4x+2y-1)dy=0$ [**Ans :** $x+2y+\log(2x+y-1)=c$]
11. $(2x+3y+4)dx-(4x+6y+5)dy=0$ [**Ans :** $7x$ –og$(14x+21y+22)=c$]
12. $(2x+9y-20)dx-(6x+2y-10)dy=0$

    [**Ans :** $x+2y-5=c(2x-y)^2=c$]
13. $(x-2y-5)dx-(2x+y-1)dy=0$ [**Ans :** $x^2-y^2-4xy+10x+2y=c$]
14. $(3x+2y-5)dx+(2x+3y-5)dy=0$ [**Ans :** $3(x^2+y^2)+4xy-10(x+y)=c$]
15. $(2x-4y+3)dy+(x-2y+1)dx=0$ [**Ans :** $\log[4(x-2y)+5]=4(x+2y)+c$]

4.5.4 Exact equations

Suppose the equation to be solved is of theform

$M(x, y)\,dx + N(x, y)\,dy = 0$ ... 1

where M and N are functions of x and y such that

$$\frac{\partial M}{\partial y} = \frac{\partial N}{\partial x} \qquad \dots 2$$

Let us put $\quad F = F(x, y) = \int M(x,y)dx$ ... 3

with y kept constant while integrating M. Then

$$\frac{\partial F}{\partial x} = M \qquad \dots 4$$

and $\frac{\partial}{\partial y}\left(\frac{\partial F}{\partial x}\right) = \frac{\partial M}{\partial y} = \frac{\partial N}{\partial x}$ by (2)

This gives $$\frac{\partial}{\partial x}\left(\frac{\partial F}{\partial y} - N\right) = 0$$

or $$\frac{\partial F}{\partial y} - N = \varphi'(y) \text{ (say)} \qquad \ldots 5$$

where $\varphi'(y)$ is a function of y.

Since $F = F(x, y)$ we have

$$dF = \frac{\partial F}{\partial x}dx + \frac{\partial F}{\partial y}dy$$

On using (4) and (5) this becomes

$$dF = Mdx + \{N + \varphi'(y)\}\, dy$$

or $$Mdx + Ndy = dF - \varphi'(y)\, dy = dF - d\varphi$$
$$= d(F - \varphi) \qquad \ldots 6$$

This shows that the left hand side of the given equation (1) is an exact differential of the function $(F - \varphi)$. For this reason, when the condition (2) holds, equation (1) is called an exact differential equation; then the equation can be rewritten as

$$d(F - \varphi) = 0$$

Integrating this, we get

$$F - \varphi = c, \text{ a constant} \qquad \ldots 7$$

this is the general solution of the equation (1)

In view of the definitions of $F(x, y)$ and $\varphi(y)$ [see (3) and (5)] the following working rule is used to write down the general solution (7)

Working rule

1. Integrate $M(x, y)$ w.r.t x treating y as constant. Let I_1 be this integral
2. Integrate w.r.t y only those terms in $N(x, y)$ that do not contain x. Let I_2 be this integral.
3. Then the sum of the above intergrals equated to a constant c; that is $I_1 + I_2 = c$, is the general solution of the given equation.

WORKED EXAMPLES

Example 1 Solve : $3x(xy - 2)\, dx + (x^3 + 2y)dy = 0$.

Solution

The given equation is of the form $Mdx + Ndy = 0$ with $M = 3x(xy - 2)$, $N = x^3 + 2y$, we find that

$$\frac{\partial M}{\partial y} = 3x^2 = \frac{\partial N}{\partial x}$$

Therefore, the given equation is exact.
Integrating M w.r.t. x, treating y as constant, we get the integral

$$I_1 = x^3y - 3x^2$$

Integrating w.r.t. y only those terms in N which do not contain x, we get the integral

$$I_2 = y^2$$

Hence, the general solution of the given equation is

$$I_1 + I_2 \equiv x^3y - 3x^2 + y^2 = \text{c, a constant.}$$

Example 2 Solve: $(x^2 + y^2 - a^2)\, xdx + (x^2 - y^2 - b^2)\, ydy = 0$

Solution
The given equation is of the form Mdx + Ndy = 0 with
$M = x\,(x^2 + y^2 - a^2)$, $N = y\,(x^2 - y^2 - b^2)$, we find that

$$\frac{\partial M}{\partial y} = 2xy = \frac{\partial N}{\partial x}$$

Therefore, the given equation is exact.
Integrating M w.r.t x, treating y as constant, we get the integral

$$I_1 = \frac{x^4}{4} + (y^2 - a^2)\frac{x^2}{2}$$

Integrating w.r.t. y only those terms in N which do not contain x, we get the integral

$$I_2 = -\frac{y^4}{4} - b^2\frac{y^2}{2}$$

Hence, the general solution of the given equation is

$$I_1 + I_2 \equiv \frac{x^4}{4} + (y^2 - a^2)\frac{x^2}{2} - \frac{y^4}{4} - b^2\frac{y^2}{2} = c\text{, a constant}$$

or $\quad x^4 + 2x^2(y^2 - a^2) - y^4 - 2b^2y^2 = 4c$

Example 3 Solve : $(3x^2y^2 + x^2)\, dx + (2x^3\, y + y^2)\, dy = 0$

Solution
The given equation is of the form $Mdx + Ndy = 0$ with
$M = 3x^2\, y^2 + x^2$, $N = 2x^3y + y^2$ we find

$$\frac{\partial M}{\partial y} = 6x^2y = \frac{\partial N}{\partial x}$$

Therefore, the given equation is exact.
Integrating M w.r.t x, treating y as constant, we get the integral

$$I_1 = x^3y^2 + \frac{x^3}{3}$$

Integrating w.r.t. y only those terms in N which do not contain x, we get the integral

$$I_2 = \frac{y^3}{3}$$

Hence, the general solution of the given equation is

$$I_1 + I_2 \equiv x^3y^2 + \frac{x^3}{3} + \frac{y^3}{3} = c \text{ a constant}$$

or $x^3 + y^3 + 3x^3y^2 = 3c$

Example 4 Solve : $(5x^4 - 3x^2y^2 - 2xy^3)\, dx + (2x^3y - 3x^2y^2 - 5y^4)dy = 0$

Solution

The given equation is of the form $Mdx + Ndy = 0$ with
$M = 5x^4 + 3x^2y^2 - 2xy^3$, $N = 2x^3y - 3x^2y^2 - 5y^4$ we find

$$\text{we find that } \frac{\partial M}{\partial y} = 6x^2y - 6xy^2 = \frac{\partial N}{\partial x}$$

Therefore, the given equation is exact.
Integrating M w.r.t x, treating y as constant, we get the integral

$$I_1 = x^5 + x^3y^2 - x^2y^3$$

Integrating w.r.t. y only those terms in N which do not contain x, we get the integral

$$I_2 = -y^5$$

Hence, the general solution of the given equation is

$$I_1 + I_2 \equiv x^5 + x^3y^2 - x^2y^3 - y^5 = c, \text{ a constant}$$

Example 5 Solve: $(y \sin 2x)\, dx - (1 + y^2 + \cos^2 x)\, dy = 0$

Solution

The given equation is of the form $Mdx + Ndy = 0$ with
$M = y \sin 2x$, $N = -(1 + y^2 + \cos^2 x)$ we find that

$$\frac{\partial M}{\partial y} = \sin 2x, \quad \frac{\partial N}{\partial x} = -(2\cos x)(-\sin x) = \sin 2x$$

$$\text{Thus} \quad \frac{\partial M}{\partial y} = \frac{\partial N}{\partial x} \text{ the given equation is exact.}$$

Integrating M w.r.t. x, treating y as constant, we get the integral

$$I_1 = -(1/2)y \cos 2x.$$

Integrating w.r.t. y only those terms in N which do not contain x, we get the integral

$$I_2 = -(y + y^3/3)$$

Hence, the general solution of the given equation is

$$I_1 + I_2 \equiv -\frac{1}{2}y\cos 2x - \left(y + \frac{1}{3}y^3\right) = \text{constant}$$

or $3y \cos 2x + 6y + 2y^3 = c$, a constant

Example 6 Solve : $\frac{dy}{dx} + \frac{y\cos + \sin y + y}{\sin x + x\cos y + x} = 0$

Solution

The given equation may be rewritten as

$(y \cos x + \sin y + y)\, dx + (\sin x + x \cos y + x)\, dy = 0$

Here $M = y\cos x + \sin y + y$, $N = \sin x + x \cos y + x$

$$\frac{\partial M}{\partial y} = \cos x + \cos y + 1 = \frac{\partial N}{\partial x}$$

Therefore, the given equation is exact.

Integrating M w.r.t x, treating y as constant, we get the integral

$$I_1 = y \sin x + x (\sin y + y)$$

We not that x is present in all terms of N. Therefore, the integral I_2 associated with N is zero.

Hence, the general solution of the given equation is

$$I_1 \equiv y \sin x + x (\sin y + y) = c$$

Example 7 Solve $\frac{dy}{dx} = \frac{x(2\log x + 1)}{\sin y + y\cos y}$

Solution

The given equation may be rewritten as

$(\sin y + y\cos y)dy = x\,(2\log x + 1)dx$

or $\quad d(y\sin y) = d(x^2 \log x)$

or $\quad d\{y \sin y - x^2\log x\} = 0$

Evidently, the general solution of this equation is

$y \sin y - x^2\log x = c$, a constant.

Example 8 Solve : $\frac{dy}{dx} = \frac{xy^2 - 1}{1 - x^2 y}$

Solution

The given equation may be rewritten as

$$(xy^2 - 1)\, dx + (x^2y - 1)dy = 0$$

or $\quad (xy^2dx + x^2ydy) - (dx + dy) = 0$

or $\quad d\left(\frac{1}{2}x^2y^2\right) - d(x + y) = 0$

or $$d\left\{\frac{1}{2}x^2y^2-(x+y)\right\} = 0$$

Evidently, the general solution of this equation is

$$\frac{1}{2}x^2y^2-(x+y) = c, \text{ a constant}$$

Example 9

Solve: $[\cos x \tan y + \cos(x+y)]dx + [\sin x \sec^2 y + \cos(x+y)]dy = 0$

Solution

The given equation may be rewritten as

$(\cos x \tan y\, dx + \sin x \sec^2 y\, dy) + \cos(x+y)(dx+dy) = 0$

or $d(\sin x \tan y) + d[\sin(x+y)] = 0$

Evidently, the general solution of this is

$\sin x \tan y + \sin(x+y) = c$, a constant.

Example 10 Solve $\left[y\left(1+\frac{1}{x}\right)+\cos y\right]dx+(x+\log x - x\sin y)dy = 0$

Solution

The given equation may be rewritten as

$(y\,dx + x\,dy)+\left(\frac{y}{x}dx + \log x\,dy\right) + (\cos y\, dx - x \sin y\, dy) = 0$

or $d(xy) + d(y \log x) + d(x \cos y) = 0$

Evidently, the general solution of this equation is

$xy + y \log x + x \cos x\, y = c$, a constant

Example 11 Solve : $x\,dx + y\,dy = \dfrac{a^2(x\,dy - y\,dx)}{x^2+y^2}$

Solution

We first note that

$$xdx + ydy = \frac{1}{2}d(x^2+y^2) \qquad \ldots \text{ i}$$

and

$$\frac{xdy - ydx}{x^2+y^2} = \frac{xdy-ydx}{x^2}\cdot\frac{1}{1+(y/x)^2}$$

$$= d\left(\frac{y}{x}\right)\cdot\frac{1}{1+(y/x)^2}$$

$$= d[\tan^{-1}(y/x)] \qquad \ldots \text{ ii}$$

In view of the relations (i) and (ii), the given equation may be rewritten as

$$d\left\{\frac{1}{2}(x^2+y^2)-a^2\tan^{-1}(y/x)\right\} = 0$$

Evidently, the general solution of the equation is

$$\frac{1}{2}(x^2+y^2)-a^2\tan^{-1}(y/x) = c$$

EXERCISES

Solve the following equations

1. $(x^2-4xy-2y^2)dx+(y^2-4xy-2x^2)dy=0$

$$\left[\textbf{Ans}: x^3+y^3-6xy(x+y)=3c\right]$$

2. $(x^2-2xy+3y^2)dx+(y^2+6xy-x^2)dy=0$

$$\left[\textbf{Ans}: x^3-3x^2y+9xy^2+y^3=c\right]$$

3. $(a^2-2xy-y^2)dx-(x+y)^2dy=0$

$$\left[\textbf{Ans}: 3a^2x-3x^2y-3xy^2-y^3=c\right]$$

4. $(x^4-2xy^2+y^4)dx-(2x^2y-4xy^3)dy=0$

$$\left[\textbf{Ans}: x^5=5(x^2y^2-xy^4+c)\right]$$

5. $(3x^2+6xy^2)dx+(6x^2y+4y^3)dy=0$ $\left[\textbf{Ans}: x^3+3x^2y^2+y^4=c\right]$
6. $(2x^2+6xy-y^2)dx+(3x^2-2xy+y^2)dy=0$

$$\left[\textbf{Ans}: x^3+9x^2y-3xy^2+y^3=c\right]$$

7. $(2x^3-xy^2-2y+3)dx-(x^2y+2x)dy=0$

$$\left[\textbf{Ans}: x^4-x^2y^2-4xy+6x=c\right]$$

8. $(2x/y^3)dx+\left\{y^2-3x^2)/y^4\right\}dy=0$ $\left[\textbf{Ans}: x^2-y^2=cy^3\right]$
9. $(e^y+1)\cos x dx+e^y\sin x dy=0$ $\left[\textbf{Ans}: (e^y+1)\sin x=c\right]$
10. $(e^y+y\cos xy)dx+(xe^y+x\cos xy)dy=0$ $\left[\textbf{Ans}: xe^y+\sin xy=c\right]$
11. $(\sec x\tan x\tan y-e^x)dx+\sec x\sec^2 y dy=0$

$$\left[\textbf{Ans}: \sec x\tan y-e^x=c\right]$$

12. $(y^2 e^{xy^2} + 4x^3)dx + (2xye^{xy^2} - 3y^2)dy = 0$

$$\left[\textbf{Ans}: e^{xy^2} + x^4 - y^3 = c\right]$$

13. $(2y\sin x + \cos y)dx - (x\sin y + 2\cos x + \tan y)dy = 0$

$$[\textbf{Ans}: x\cos y - 2y\cos x - \log\sec y = c]$$

14. $(2xy + y - \tan y)dx + (x^2 - x\tan^2 y + \sec^2 y)dy = 0$

$$\left[\textbf{Ans}: x^2 y + xy - x\tan y + \tan y = c\right]$$

15. $\dfrac{dy}{dx} = \dfrac{x^2 - 2xy}{x^2 - \sin y}$ $\quad\left[\textbf{Ans}: x^3 - 3(x^2 y + \cos y) = c\right]$

16. $\dfrac{dy}{dx} = \dfrac{\cos x - 3x^2 \tan y}{x^3 \sec^2 y}$ $\quad\left[\textbf{Ans}: x^3 \tan y - \sin x = c\right]$

4.5.5 Equations reducible to exact equations

May differential equations of the form $Mdx + Ndy = 0$ that are not *exact* can be made exact by multiplying them by appropriate functions called *integrating factors*. The following rules are followed for finding the integrating factos.

Rules for integrating factors

1. If the equation is homogeneous and $Mx + Ny \neq 0$, then $\dfrac{1}{Mx + Ny}$ is an integrating factor.
2. If $\dfrac{1}{N}\left(\dfrac{\partial M}{\partial y} - \dfrac{\partial N}{\partial x}\right)$ is a function of x alone, say $f(x)$ then $e^{\int f(x)dx}$ is an integrating factor.
3. If $\dfrac{1}{M}\left(\dfrac{\partial N}{\partial x} - \dfrac{\partial M}{\partial y}\right)$ is a function of y alone, say $g(y)$, then $e^{\int g(y)dy}$ is an integrating factor.
4. If the equation is of the form
   $f_1(x, y)ydx + f_2(x, y)xdy = 0$
   so that $M = y f_1(x, y)$ and $N = xf_2(x, y)$ and if $Mx - Ny \neq 0$ then $\dfrac{1}{Mx - Ny}$ is an integrating factor.
5. If the equation is of the form
   $x^m y^n (ay\, dx + bx\, dy) + x^p y^q (cydx + exdy) = 0$

where *m, n, p, q, a, b, c,* e are all constants, then xh yh is an integrating factor. Here *h* and *k* are constants to be chosen so that the condition for exactness is satisfied.

Example 1 Solve: $(x^2y - 2xy^2)\,dx - (x^3 - 3x^2y)\,dy = 0$

Solution

The given equation is of the form $Mdx + Ndy = 0$ with
$M = x^2y - 2xy^2, \quad N = -(x^3 - 3x^2y)$

We find that $\dfrac{\partial M}{\partial y} \neq \dfrac{\partial N}{\partial x}$. Therefore the equation is not exact. But, the equation is homogeneous, and we check that $Mx + Ny = x^2y^2 \neq 0$. Therefore, by Rule (1).

$$\frac{1}{Mx + Ny} = \frac{1}{x^2y^2}$$

is an integrating factor of the equation. Multiplying the equation by this factor, the equation becomes the *exact equation.*

$$\left(\frac{1}{y} - \frac{2}{x}\right)dx = \left(\frac{x}{y^2} - \frac{3}{y}\right)dy = 0$$

Solving this exact equation, we get the integral

$$\frac{1}{y}\int dx - 2\int \frac{dx}{x} + 3 + \int \frac{dy}{y}\,dy = c, \text{ a constant}$$

i.e.
$$\frac{x}{y} - 2\log x + 3\log y = c$$

Example 2 Solve: $(x^2 - 3xy + 2y^2)\,dx + x(3x - 2y)\,dy = 0$

Solution

Here $M = x^2 - 3xy + 2y^2$ and $N = x(3x - 2y)$. We veriefy that $\dfrac{\partial M}{\partial y} \neq \dfrac{\partial N}{\partial x}$. Therefore the equation is not exact. But, the equation is homogeneous, and we find that $Mx + Ny = x^3 \neq 0$.

Therefore,

$$\frac{1}{Mx + Ny} = \frac{1}{x^3}$$

is an integrating factor for the given equation. Multiplying the equation by this factor, the equation becomes the exact equation:

$$\left(\frac{1}{x} - \frac{3}{x^2}y + 2\frac{y^2}{x^3}\right)dx + \left(\frac{3}{x} - \frac{2y}{x^2}\right)dy = 0$$

The general solution of this exact equation, which is of the form $Md + Ndy = 0$, is obtained, as usual, by integrating M w.r.t. x, treating y as constant and integrating w.r.t. y only those terms in N that do not contain x, and finally equating the sum of the integrals to a constnat. Thus, we get

$$\log x + \frac{3y}{x} - \frac{y^2}{x^2} = c$$

This is the general solution fo the given equation.

Example 3 Solve: $(x^2 + y^3 + 6x)\, dx + y^2 x\, dy = 0$

Solution

The given equation is of the form $Mdx + Ndy = 0$ with

$$M = x^2 + y^3 + 6x, \quad N = y^2 x$$

We find that

$$\frac{\partial M}{\partial y} - \frac{\partial N}{\partial x} = 3y^2 - y^2 = 2y^2 \neq 0$$

so that the equation is not exact. We find that

$$\frac{1}{N}\left(\frac{\partial M}{\partial y} - \frac{\partial N}{\partial x}\right) = \frac{2y^2}{y^2 x} = \frac{2}{x} = f(x), \text{ say}$$

Therefore, by Rule (2), an integrating factor for the given equation is

$$e^{\int f(x)dx} = e^{\int (2/x)dx} = e^{2\log x} = x^2$$

Multiplying the equation by this factor, we get the exact equation.

$$(x^4 + x^2y^3 + 6x^3)\, dx + y^2 x^3\, dy = 0$$

The general solution of this exact equation is

$$\int x^4 dx + y^3 \int x^2 dx + 6\int x^3 dx = c, \text{ a constant}$$

i.e.,

$$\frac{x^5}{5} + \frac{y^3 x^3}{3} + \frac{6x^4}{4} = c$$

or

$$6x^5 + 10x^3y^3 + 45x^4 = 30c.$$

Example 4 Solve: $(6x^2 + 4y^3 + 12y)\, dx + 3x\,(1 + y^2)\, dy = 0$

Solution

Here $M = 6x^2 + 4y^3 + 12y$, $N = 3x(1 + y^2)$, so that

$$\frac{\partial M}{\partial y} - \frac{\partial N}{\partial x} = 12\,(y^2 + 1) - 3\,(1 + y^2) = 9\,(1 + y^2)$$

This yields, $\frac{1}{N}\left(\frac{\partial M}{\partial y} - \frac{\partial N}{\partial x}\right) = \frac{3}{x} = f(x)$, say

Therefore, an integrating factor for the given equation is

$$e^{\int f(x)dx} = e^{3\int \frac{1}{x}dx} = e^{3\log x} = e^{\log x^3} = x^3$$

Multiplying the equation by this factor, we get the equation

$$(6x^5 + 4x^3y^3 + 12x^3y)\, dx + 3x^4 (1 + y^2)\, dy = 0$$

The general solution of this exact equation is

$$x^6 + x^4 y^3 + 3yx^4 = c$$

Example 5 Solve: $(y^4 + 2y)\, dx + (xy^3 + 2y^4 - 4x)\, dy = 0$

Solution

Here $M = y^4 + 2y$, $N = xy^3 + 2y^4 - 4x$, so that

$$\frac{\partial M}{\partial y} - \frac{\partial N}{\partial x} = 4y^3 + 2 - (y^3 - 4) = 3\,(y^3 + 2) \neq 0$$

so that the equation is not exact, We find

$$\frac{1}{M}\left(\frac{\partial N}{\partial x} - \frac{\partial M}{\partial y}\right) = \frac{3\left(y^3 + 2\right)}{y^4 + 2y} = -\frac{3}{y} = g(y)\text{, say}$$

Therefore,

$$e^{\int g(y)dx} = e^{-\int (3/y)dx} = e^{-3\log y} = \frac{1}{y^3}$$

is an integrating factor of the given equation. Multiplying by this factor, the equation becomes the exact equation.

$$\left(y + \frac{2}{y^2}\right)dx + \left(x + 2y - 4\frac{x}{y^3}\right)dy = 0$$

The general solution of this exact equation is

$$\left(y + \frac{2}{y^2}\right)x + y^2 = c$$

Example 7 Solve: $(x^2y^2 + xy + 1)y\, dx + (x^2y^2 - xy + 1)\, x\, dy = 0$

Solution

The given equation is of the form $Mdx + Ndy = 0$, with

$$M = (x^2y^2 + xy + 1)y = y f_1(x, y), \text{ say}$$

and

$$N = (x^2y^2 - xy + 1)y = x f_2(x, y), \text{ say}$$

We check that $Mx - Ny = 2x^2y^2 \neq 0$. Therefore, by Rule (4)

$$\frac{1}{Mx - Ny} = \frac{1}{2x^2y^2}$$

is an integrating factor for the given equation.

Multiplying by this factor, the equation becomes the exact equation.

$$\left(1+\frac{1}{xy}+\frac{1}{x^2y^2}\right)y\,dx+\left(1-\frac{1}{xy}+\frac{1}{x^2y^2}\right)x\,dy = 0$$

The general solution of this exact equation is

$$y\int dx+\int\frac{1}{x}dx+\frac{1}{y}\int\frac{dx}{x^2}-\int\frac{1}{y}dy = \text{c}$$

i.e.

$$yx+\log x-\frac{1}{yx}-\log y = c$$

where c is a constant.

Example 8 Solve : $(xy\sin xy+\cos xy)\,y\,dx+(xy\sin xy-\cos xy)\,x\,dy=0$

Solution

Here, $\quad \text{M}=(xy\sin xy+\cos xy)\,y, \quad \text{N}=(xy\sin xy-\cos xy)\,x$

$$Mx-Ny = 2xy\cos xy \neq 0$$

Therefore, $\dfrac{1}{Mx-Ny}=\dfrac{1}{2xy\cos xy}$ is an integrating factor for the given equation. Multiplying the equation by this factor, we get

$$\left(y+\tan xy+\frac{1}{x}\right)dx+\left(x\tan xy-\frac{1}{y}\right)dy=0$$

The general solution of this exact equation is

$$\log\sec xy+\log x-\log y = \log c,$$

or

$$x\sec xy = cy$$

Example 9 Solve : $(y^2+2x^2y)\,dx+(2x^3-xy)\,dy=0$

Solution

The given equation may be rewritten as

$$y(y\,dx-x\,dy)+2x^2\,(y\,dx+x\,dy) = 0 \qquad \dots i$$

This form of the equation suggest that by rule (5) $x^h\,y^h$ is an integrating factor, where h and k are constants to be determined. Multiplying the equation by this factor, we get

$$x^h y^k\,(y^2+2x^2\,y)\,dx+x^h\,y^k\,(2x^3-xy)\,dy = 0 \qquad \dots ii$$

Here

$$M = x^h\,y^{k+2}+2x^{h+2}\,y^{k+1}$$
$$N = 2x^{h+3}\,y^h-x^{h+1}\,y^{h+1}$$

so that

$$\frac{\partial M}{\partial y} = (h+2)x^h y^{k+1} + 2(k+1)x^{h+2}y^k$$

$$\frac{\partial N}{\partial x} = 2(h+3)x^{h+2}y^k - (h+1)x^h y^{k+1}$$

Therefore, for the equation (ii) the condition $\frac{\partial M}{\partial y} = \frac{\partial N}{\partial x}$ holds if

$$(k+2)y + 2(k+1)x^2 = 2(h+3)x^2 - (h+1)y$$

or, on equating the corresponding coefficients,

$$k+2 = -(h+1), \quad k+1 = h+3$$

or $\qquad h+k+3 = 0, \quad h-k+2 = 0$

Solving these, we get $h = -5/2, \ k = -1/2$

Thus for $h = -5/2$ and $k = -1/2$, the equation (ii) is exact. Putting these values of h and k into (ii), we get the equation

$$(x^{-5/2}y^{3/2} + 2x^{-1/2}y^{1/2})dx + (2x^{1/2}y^{-1/2} - x^{-3/2}y^{1/2})dy = 0$$

The general solution of thie exact equation is found as

$6x^{1/2}y^{1/2} - x^{-3/2}y^{3/2} = c$

Example 10 Solve: $xy^3(ydx + 2xdy) + (3ydx + 5x\,dy) = 0$

Solution

This form of the given equation suggest that $x^h y^k$ is an integrating factor, where h and k are constants to be determined. Multiplying by this factor, the equation becomes

$x^h y^h [(xy^4 + 3y)\,dx + (2x^2y^3 + 5x)\,dy] = 0 \qquad \ldots ii$

which is of the form $Mdx + Ndy = 0$, where

$$M = x^h y^k(xy^4 + 3y) = x^{h+1}y^{k+4} + 3x^h y^{k+1}$$

and $\qquad N = x^h y^k(2x^2y^3 + 5x) = 2x^{h+2}y^{k+3} + 5x^{h+1}y^k$

These yield

$$\frac{\partial M}{\partial y} = (k+4)x^{h+1}y^{k+3} + 3(k+1)x^h y^k$$

$$\frac{\partial N}{\partial x} = 2(h+2)x^{h+1}y^{k+2} + 5(h+1)x^h y^k$$

Now, we find that the condition $\frac{\partial M}{\partial y} = \frac{\partial N}{\partial x}$ holds if

$$(k+4)xy^3 + 3(k+1) = 2(h+2)xy^3 + 5(h+1)$$

Equating the correspoding coefficients in this, we get

$$k+4=2(h+2) \text{ and } 3(h+1)=5(h+1)$$

or $h=2k$, and $3k-5h=2$

which yield, no solving, $h=2$, $k=4$

Thus, for $h=2$ and $k=4$, the equation (i) is exact.

Putting these values of h and k into (i), we get the exact equation.

$$(x^3y^8 + 3x^2y^5)\,dx + (2x^4y^7 + 5x^3y^4)\,dy = 0$$

The general solution of thie exact equation is

$$y^8 \int x^3 dx + 3y^5 \int x^2 dx = c, \text{ a constant}$$

i.e.,
$$x^4y^8 + 4x^3y^5 = 4c.$$

EXERCISES

Solve the following differential equations

1. $x^2ydx-(x^3+y^3)dy=0$ $\left[\textbf{Ans}: x^3/3y^3-\log y=c\right]$
2. $(xy^3+y)dx+2(x^2y^2+x+y^4)dy=0$

   $\left[\textbf{Ans}: 3x^2y^4+6xy^2+2y^6=c\right]$
3. $(8xy-9y^2)dx+2(x^2-3xy)dy=0$ $\left[\textbf{Ans}: x^3y(2x-3y)=c\right]$
4. $(1+xy)y\,dx+(1-xy)xdy=0$ $[\textbf{Ans}: \log(x/y)-1/xy)=c]$
5. $y(xy+2x^2y^2)dx+x(xy-x^2y^2)dy=0$

   $[\textbf{Ans}: (1/xy)=c+\log(x^2/y)]$
6. $xdy-ydx+a(x^2+y^2)dx=0$ $[\textbf{Ans}: ax+\tan^{-1}(y/x)=c]$
7. $ydx-xdy+\log xdx=0$ $[\textbf{Ans}: y+cx+\log x+1=0]$
8. $ydx-xdy-3x^2y^2e^{x^3}dx=0$ $[\textbf{Ans}: x/y-e^{x^3}=c]$

4.5.6 Linear equations

Suppose the equation to be solved is of the form

$$\frac{dy}{dx}+P(x)y = Q(x) \qquad \dots 1$$

where P and Q are functions of Such an equation is called a *linear differential equation*. (Also known as Leibniz's linear differential equation)

Let us take

$$R = R(x)=e^{\int P dx} \qquad \dots 2$$

Then we find that

$$\frac{dR}{dx} = e^{\int P dx}\frac{d}{dx}\left[\int Pdx\right] = RP \qquad \dots 3$$

and

$$\frac{d}{dx}(Ry) = R\frac{dy}{dx} + y\frac{dR}{dx} = R\left[\frac{dy}{dx} + Py\right] \qquad \dots 4$$

using (3). Hence, on multiplying by R, equation (1) takes the form

$$\frac{d}{dx}(Ry) = RQ$$

Integrating this, we get

$$Ry = \int RQdx + c$$

where c is a constant. Using (2) this reads

$$ye^{\int P dx} = \int Q(x)e^{\int P dx}dx + c \qquad \dots 5$$

This is the general solution of equation (1)

Note: The function $R = R(x) = e^{\int P dx}$ is called an ***integrating factor (I.F)*** of of equation (1).

WORKED EXAMPLES

Example 1 Solve: $\dfrac{dy}{dx} + y\sec x = \tan x$

Solution The given equation is of the form (1) with

$$P = \sec x, \quad Q = \tan x$$

Therefore, $\int Pdx = \int \sec x = \log(\sec x + \tan x)$

So that $\qquad I.F. = e^{\int P dx} = \sec x + \tan x$

Hence the general solution of the given equation is [by formula (5)]

$$(\sec x + \tan x) = \int(\tan x)(\sec x + \tan x)dx + c$$

$$y(\sec x + \tan x) = \int \sec x \tan x dx + \int(\sec^2 x - 1)dx + c$$

or

$$y = 1 + \frac{c - x}{\sec x + \tan x}$$

Example 2 Solve $\dfrac{dy}{dx} + y\cot x = \cos x$

Solution The given equation is a linear differential equation with $P = \cot x$, $Q = \cos x$

Therefore, $\int P\,dx = \int \cot x = \log \sin x$

So that $\quad I.F = e^{\int P\,dx} = \sin x$

Hence the general solution of the given equation is

$$y \sin x = \int \cos x \sin x\,dx + c = \frac{1}{2}\sin^2 x\,dx + c$$

or $\quad y = \frac{1}{2}\sin x + c \operatorname{cosec} x$

where c is a constant.

Example 3 Solve: $x \log x \frac{dy}{dx} + y = \log x$

Solution Dividing throughout by $x \log x$, the given equation becomes

$$\frac{dy}{dx} + \frac{1}{x \log x} y = \frac{2}{x}$$

Which is a linear differential equation with

$$P = \frac{1}{x \log x}, \; Q = \frac{2}{x}$$

Therefore, $\quad \int P\,dx = \int \frac{1/x}{\log x} = \log \log x$

So that $\quad I.F = e^{\int P\,dx} = \log x$

Hence, the general solution of the given equation is

$$y \log x = \int \frac{2}{x} \log x\,dx + c$$

$$= 2 \cdot \int \log x \left(\frac{1}{x}dx\right) + c = 2 \cdot \frac{(\log x)^2}{2} + c$$

or $\quad y = \log x + \frac{c}{\log x}$

Example 4 Solve: $\sin^2 x \frac{dy}{dx} - y = \cot x$

Solution Dividing by $\sin^2 x$, the given equation becomes

$$\frac{dy}{dx} - (\operatorname{cosec}^2 x) y = \cot x \operatorname{cosec}^2 x \qquad \ldots \text{i}$$

This is a linear differential equation with

$P = -\operatorname{cosec}^2 x, \quad Q = \cot x \operatorname{cosec}^2 x.$

Therefore, $\int P\,dx = -\int \text{cosec}^2 x = \cot x$

So that $I.f. = e^{\int P\,dx} = e^{\cot x}$

Accordingly, the general solution of the given equation is

$$ye^{\cot x} = \int(\cot x\,\text{cosec}^2 x)e^{\cot x}\,dx + c \qquad \ldots ii$$

To evaluate the integral, we put $\cot x = t$ so that $-\text{cosec}^2 x\,dx = dt$

Hence $\int(\cot x\,\text{cosec}^2 x)e^{\cot x}\,dx = -\int te^t\,dt = -\left[te^t - \int e^t\,dt\right]$

$$= e^t(1-t) = e^{\cot x}(1-\cot x)$$

Using this in (ii), we get the general solution of the given equation as

$$y = (1 - \cot x) + c\,e^{-\cot x}$$

Example 5 Solve $x(1-x^2)\dfrac{dy}{dx} + \left(2x^2 - 1\right)y = x^3$

Solution The given equation may be rewritten as

$$\frac{dy}{dx} + \frac{2x^2-1}{x(1-x^2)}y = \frac{x^2}{1-x^2}$$

Which is a linear differential equation with

$$P = \frac{2x^2-1}{x(1-x^2)},\ Q = \frac{x^2}{1-x^2}$$

Setting $$\frac{2x^2-1}{x(1-x^2)} = \frac{A}{x} + \frac{B}{1-x} + \frac{C}{1+x}$$

We find $2x^2 - 1 = A(1-x^2) + Bx(1+x) + Cx(1-x)$

For $x = 0, 1, -1$ this yields, respectively,

$-1 = A,\ 1 = 2B,\ 1 = -2C$

$$\therefore \quad A = -1,\ B = 1/2,\ C = -1/2$$

$$P = \frac{2x^2-1}{x(1-x^2)} = -\frac{1}{x} + \frac{1}{2(1-x)} - \frac{1}{2(1+x)}$$

Therefore,

$$\int P\,dx = -\log x - \frac{1}{2}\log(1-x) - \frac{1}{2}\log(1+x)$$

$$= -\log x\sqrt{1-x^2}$$

So that $$\text{I.F} = e^{\int P\,dx} = \frac{1}{x\sqrt{1-x^2}}$$

Hence, the general solution of the given equation is

$$y \cdot \frac{1}{x\sqrt{1-x^2}} = \int \frac{x^2}{1-x^2} \cdot \frac{1}{x\sqrt{1-x^2}} dx + c$$

$$= \int \frac{x}{(1-x^2)^{3/2}} dx + c$$

$$= -\frac{1}{2}\int \frac{dt}{t^{3/2}} + c, \text{ where } t = 1 - x^2$$

$$= \frac{1}{\sqrt{t}} + c = \frac{1}{\sqrt{1-x^2}} + c$$

$$y = x + cx\sqrt{1-x^2}$$

Here c is a constant.

Example 6 Slove $dr + (2r\cot\theta + \sin 2\theta)d\theta = 0$

Solution The given equation may be rewritten as

$$\frac{dr}{d\theta} + (2\cot\theta)r = -\sin 2\theta \qquad \ldots 1$$

which is a linear differential equation with θ as an independent variable and r as the dependent variable. For this equation,

$$\text{I.F} = e^{\int(2\cot\theta)d\theta} = e^{2\log\sin\theta} = \sin^2\theta$$

Therefore , the general solution of the equation is

$$r\sin^2\theta = \int(-\sin 2\theta)\sin^2\theta d\theta + c$$

$$= -2\int\sin^3\theta\cos\theta d\theta + c = -\frac{1}{2}\sin^4\theta + c$$

or $2r\sin^2\theta + \sin^4\theta = 2c$

Example 7 Solve $(x + \tan y)dy = \sin 2y dx$

Solution The given equation may be rewritten as

$$\frac{dx}{dy} - \frac{1}{\sin 2y}x = \frac{\tan y}{\sin 2y} \qquad \ldots \text{i}$$

This is similar to equation (1) with the roles of x and y interchanged. That is (i) is a linear differential equation which y is the independent variable and x as the dependent variable. For this equation,

$$P = -\frac{1}{\sin 2y} = -\operatorname{cosec} 2y, \quad Q = \frac{\tan y}{\sin 2y} = \frac{1}{2}\sec^2 y$$

Therefore,

$$\int P\,dy = -\int \operatorname{cosec} 2y\,dy = -\frac{1}{2}\log\tan y = \log\left(\frac{1}{\sqrt{\tan y}}\right)$$

So that $$\text{I.F} = e^{\int P\,dy} = e^{\int \log\left(\frac{1}{\sqrt{\tan y}}\right)dy} = \frac{1}{\sqrt{\tan y}}$$

Hence, the general solution of the equation (i) is

$$xe^{\int P\,dy} = \int Qe^{\int P\,dy}\,dy + c$$

$$\frac{x}{\sqrt{\tan y}} = \frac{1}{2}\int \sec^2 y \frac{1}{\sqrt{\tan y}}\,dy + c$$

$$= \sqrt{\tan y} + c$$

$$x = \tan y + c\sqrt{\tan y}$$

Example 8 Solve $ye^y dx = (y^3 + 2xe^y)dy$

Solution The given equation may be rewritten as

$$\frac{dx}{dy} - \frac{2}{y}x = y^2 e^{-y}$$

That is a linear differential equation with y as an independent variable and x as the dependent variable. For this equation,

$$P = -\frac{2}{y}, \quad Q = y^2 e^{-y}$$

Therefore,

$$\int P\,dy = -\int \frac{2}{y}\,dy = -2\log y$$

So that $$I.F = e^{\int P\,dy} = e^{-2\log y} = \frac{1}{y^2}$$

Hence, the general solution of the equation is

$$xe^{\int P\,dy} = \int Qe^{\int P\,dy}\,dy + c$$

$$\frac{x}{y^2} = \int \left(y^2 e^{-y}\right)\frac{1}{y^2}\,dy + c$$

$$= \int (e^{-y})\,dy + c$$

$$= -e^{-y} + c$$

$$x = (-e^{-y} + c)y^2$$

Example 9 Solve $(1+y^2)dx = (\tan^{-1} y - x)dy$

Solution The given equation may be rewritten as

$$\frac{dx}{dy}(1+y^2)+x = \tan^{-1}y$$

$$\frac{dx}{dy}+\frac{1}{(1+y^2)}\cdot x = \frac{\tan^{-1} y}{1+y^2}$$

That is a linear differential equation which y is an independent variable and x is the dependent variable. For this equation,

$$P=\frac{1}{1+y^2}, \quad Q=\frac{\tan^{-1} y}{1+y^2}$$

Therefore,

$$\int P\,dy = \int\frac{1}{1+y^2}dy = \tan^{-1} y$$

So that $\quad$ I.F $= e^{\int P\,dy} = e^{\tan^{-1} y}$

Hence, the general solution of the equation is

$$xe^{\int P\,dy} = \int Qe^{\int P\,dy}dy+c$$

$$xe^{\tan^{-1} y} = \int\frac{\tan^{-1} y}{1+y^2}e^{\tan^{-1} y}dy+c$$

$$= \int(te^t)dt+c, \quad \text{where } t=\tan^{-1}y$$

$$= (t-1)e^t+c$$

$$= (\tan^{-1} y-1)e^{\tan^{-1} y}+c$$

$$x = (\tan^{-1} y-1)+c\,e^{\tan^{-1} y}$$

Example 10 Solve $(x+y+1)\frac{dy}{dx}=1$

Solution The given equation may be rewritten as

$$\frac{dx}{dy}-x = y+1$$

which is a linear differential equation with y as an independent variable and x as the dependent variable. For this equation, $P=-1$, $Q=y+1$

Therefore,

$$\int P\,dy = -\int 1dy = -y$$

So that, $\quad$ I.F $= e^{\int P\,dy} = e^{-y}$

Hence, the general solution of the equation is

$$xe^{\int P dy} = \int Qe^{\int P dy} dy + c$$

$$xe^{-y} = \int (y+1)e^{-y} dy + c$$

where c is a constant. Carrying out the integration by parts, we get

$$xe^{y} = (y+1)(-e^{-y}) - \int (-e^{-y}) dy + c$$

$$= -(y+1)e^{-y} - e^{-y} + c$$

$$x = -(y+1) - 1 + c\, e^{y}$$

$$(x + y + 2) = c\, e^{y}$$

The equations considered in Example 11 to 15 are of the form

$$f'(y)\frac{dy}{dx} + f(y)P(x) = Q(x)$$

To solve such equations, we employ the substitution $f(y) = t$

Example 11 Solve $e^{y}\left(\frac{dy}{dx} + 1\right) = e^{x}$

The given equation may be rewritten as

$$e^{y}\frac{dy}{dx} + e^{y} = e^{x}$$

Setting $e^{y} = t$, so that $e^{y}\frac{dy}{dx} = \frac{dt}{dx}$, this equation becomes

$$\frac{dt}{dx} + t = e^{x} \qquad \ldots i$$

which is a linear differential equation with t as the dependent variable. In this equation, $P = 1,\ \ Q = e^{x}$

Therefore,

$$\int P\, dx = \int 1 dx = x$$

So that $\quad \text{I.F} = e^{\int P dx} = e^{x}$

Hence, the general solution of the equation is

$$te^{\int P dy} = \int Qe^{\int P dy} dy + c$$

$$t\, e^{x} = \int e^{x} e^{x} dx + c$$

$$= \tfrac{1}{2}e^{2x} + c$$

$$\therefore \quad t = \tfrac{1}{2}e^{x} + ce^{-x} \qquad \ldots ii$$

where c is a constant. Recalling that $t = e^{y}$, this becomes

$$e^{y} = \tfrac{1}{2}e^{x} + ce^{-x}$$

This is the general solution of the given equation.

Example 12 Solve $\frac{dy}{dx} + x\sin 2y = x^3 \cos^2 y$

Solution Dividing the equation by $\cos^2 y$, we get

$$\sec^2 y\frac{dy}{dx} + 2x\tan y = x^3 \qquad \ldots \text{ i}$$

Let us put $\tan y = t$, Then $\sec^2 y\frac{dy}{dx} = \frac{dt}{dx}$, this equation (i) becomes

$$\frac{dt}{dx} + 2xt = x^3 \qquad \ldots \textit{ ii}$$

which is a linear differential equation in which x is the independent variable and t as the dependent variable. In this equation,

$P = 2x,\ Q = x^3$

Therefore,

$$\int P\, dx = \int 2x\, dx = x^2$$

So that, $\quad$ I.F. $= e^{\int P dx} = e^{x^2}$

Hence, the general solution of (ii) is

$$te^{\int P dy} = \int Q e^{\int P dy} dy + c$$

$$te^{x^2} = \int x^3 e^{x^2} dx + c$$

$$= \frac{1}{2}\int u e^u du + c, \quad \text{where } u = x^2$$

$$= \frac{1}{2}(u-1)e^u + c$$

$$= \frac{1}{2}(x^2-1)e^{x^2} + c$$

where c is a constant. Recalling that $t = \tan y$, this becomes

$$\tan y e^{x^2} = \frac{1}{2}\left(x^2-1\right)e^{x^2} + c$$

or

$$\tan y = \frac{1}{2}\left(x^2-1\right) + ce^{-x^2}$$

Example 13 Solve $\tan y\frac{dy}{dx} + \tan x = \cos y \cos^3 x$

Solution Dividing the equation by $\cos y$, we get

$$(\sec y \tan y)\frac{dy}{dx} + \tan x \sec y = \cos^3 x \qquad \ldots \text{ i}$$

Let us put sec $y = t$. Then $(\sec y \tan y)\dfrac{dy}{dx} = \dfrac{dt}{dx}$, this equation becomes

$$\frac{dt}{dx} + (\tan x)t = \cos^3 x \qquad \ldots iii$$

which is a linear differential equation in which is the independent variable x and t as the dependent variable. In this equation,

$P = \tan x, \quad Q = \cos^3 x$

Therefore,

$$\int P\, dx = \int \tan x\, dx = \log \sec x$$

So that, $\quad I.F = e^{\int P\, dx} = e^{\log \sec x} = \sec x$

Hence, the general solution of (ii) is

$$t\, e^{\int P\, dy} = \int Q\, e^{\int P\, dy} dy + c$$

$$t \sec x = \int \cos^3 x \sec x\, dx + c$$

$$= \int \cos^2 x\, dx + c, \text{ Where } \cos^2 x = \frac{1}{2}(1 + \cos 2x)$$

$$= \frac{1}{2}\int (1 + \cos 2x) + c$$

$$= \frac{1}{2}\left[x + \frac{\sin 2x}{2}\right] + c \qquad \ldots iii$$

where c is a constant. Recalling that $t = \sec y$, this becomes

$$\sec y \sec x = \frac{1}{2}\left[x + \frac{\sin 2x}{2}\right] + c$$

$$\sec y \sec x = \frac{x}{2} + \frac{\sin 2x}{4} + c$$

This is the general solution of the given equation.

Example 14 Solve: $\dfrac{dy}{dx} - \dfrac{\tan y}{1+x} = (1+x)e^x \sec y$

Solution Dividing the equation by sec y, we get

$$\cos y \frac{dy}{dx} - \frac{\sin y}{1+x} = (1+x)e^x \qquad \ldots i$$

Let us put $\sin y = t$,

Then $\cos y \dfrac{dy}{dx} = \dfrac{dt}{dx}$ this equation (i) becomes

$$\frac{dt}{dx} - \frac{t}{1+x} = (1+x)e^x \qquad \ldots ii$$

This is a linear differential equation in which x is the independent variable and t as the dependent variable. In this equation,

$$P = -\frac{1}{1+x}, \quad Q = (1+x)e^x$$

Therefore,

$$\int P\,dx = -\int \frac{1}{1+x}dx = -\log(1+x) = \log\left(\frac{1}{1+x}\right)$$

So that $$I.F = e^{\int P\,dx} = e^{\log\left(\frac{1}{1+x}\right)} = \frac{1}{1+x}$$

Hence, the general solution of (ii) is

$$t e^{\int P\,dy} = \int Q e^{\int P\,dy} dy + c$$

$$t\frac{1}{1+x} = \int (1+x)e^x \frac{1}{1+x}dx + c$$

$$= \int e^x dx + c$$

$$t\frac{1}{1+x} = e^x + c$$

where c is a constant. Recalling that $t = \sin y$, this becomes

$$\sin y \frac{1}{1+x} = e^x + c$$

or $$\sin y = (1+x)(e^x + c)$$

This is the general solution of the given equation.

Example 15 Solve: $\frac{dy}{dx} + (2x\tan^{-1} y - x^3)(1+y^2) = 0$

Solution The given equation may be rewritten as

$$\frac{1}{1+y^2}\frac{dy}{dx} + 2x(\tan^{-1} y) = x^3 \qquad \ldots i$$

Let us put $\tan^{-1}y = t$, Then $\frac{1}{1+y^2}\frac{dy}{dx} = \frac{dt}{dx}$, equation (i) becomes

$$\frac{dt}{dx} + 2xt = x^3 \qquad \ldots ii$$

which is a linear differential equation in which x is the independent variable and t as the dependent variable. In this equation, $P = 2x$, $Q = x^3$

Therefore,

$$\int P\,dx = \int 2x\,dx = x^2$$

So that $$\text{I.F.} = e^{\int P\,dx} = e^{x^2}$$

Hence, the general solution of (*ii*) is

$$t e^{\int P\,dy} = \int Q e^{\int P\,dy} dy + c$$

$$t e^{x^2} = \int x^3 e^{x^2} dx + c$$

$$= \frac{1}{2}\int u e^u du + c, \text{ where } u = x^2$$

$$= \frac{1}{2}(u-1)e^u + c \qquad \dots ii$$

$$t e^{x^2} = \frac{1}{2}(x^2-1)e^{x^2} + c$$

where c is a constant. Recalling that $t = \tan^{-1}y$, this becomes

$$\tan^{-1} y e^{x^2} = \frac{1}{2}(x^2-1)e^{x^2} + c$$

or $$\tan^{-1}y = \frac{1}{2}(x^2-1) + ce^{-x^2}$$

This is the general solution of the given equation.

EXERCISES

Solve the following equations

1. $\dfrac{dy}{dx} + y\tan x = \sec x$ $\quad$ $\left[\textbf{Ans:}\ y = (\tan x + c)\cos x\right]$

2. $\dfrac{dy}{dx} + y\tan x = \sin 2x$ $\quad$ $\left[\textbf{Ans:}\ y = (c - 2\cos x)\cos x\right]$

3. $\dfrac{dy}{dx} + y\cot x = \sin x$ $\quad$ $\left[\textbf{Ans:}\ y\sin x = \dfrac{1}{2}\left\{x - \dfrac{1}{2}\sin 2x\right\} + c\right]$

4. $\dfrac{dy}{dx} + y\cot x = 4x\operatorname{cosec} x$ $\quad$ $\left[\textbf{Ans:}\ y\sin x = 2x^2 + c\right]$

5. $\dfrac{dy}{dx} + 3x^2 y + x^5 e^{x^3}$ $\quad$ $\left[\textbf{Ans:}\ 12y = e^{x^3}(2x^3 - 1) + ce^{-x^3}\right]$

6. $\dfrac{dy}{dx} = x^3 - 2xy$ if $y(1) = 2$ $\quad$ $\left[\textbf{Ans:}\ 2y - x^2 + 1 = 4e^{(1-x^2)}\right]$

7. $\dfrac{dy}{dx} - \dfrac{2y}{x+1} = (x+1)^3$ $\quad$ $\left[\textbf{Ans:}\ y = \dfrac{1}{2}(x+1)^4 + c(x+1)^2\right]$

8. $\dfrac{dy}{dx}+\dfrac{3x^2}{1+x^3}y=\dfrac{sin^2 x}{1+x^3}$ $\left[\textbf{Ans: } y(1+x^3)=\dfrac{1}{2}x-\dfrac{1}{4}\sin 2x+c\right]$

9. $\cos^2 x\dfrac{dy}{dx}+y=\tan x$ $\left[\textbf{Ans: } y+1=\tan x+ce^{-\tan x}\right]$

10. $\left(e^{-2\sqrt{x}}-y\right)dx=\sqrt{x}dy$ $\left[\textbf{Ans: } ye^{2\sqrt{x}}=2\sqrt{x}+c\right]$

11. $x^2\dfrac{dy}{dx}=3x^2-2xy+1$ $\left[\textbf{Ans: } y=x+\dfrac{1}{x}+\dfrac{c}{x^2}\right]$

12. $\left(x^2-1\right)\dfrac{dy}{dx}+2xy=1$ $\left[\textbf{Ans: } y(1-x^2)=c-x\right]$

13. $(x+1)\dfrac{dy}{dx}-y=e^{3x}(x+1)^2$ $\left[\textbf{Ans: } y=(x+1)\dfrac{1}{3}e^{3x}+c\right]$

14. $(x+2y^3)\dfrac{dy}{dx}=y$ $\left[\textbf{Ans: } x=y^3+cy\right]$

15. $dx+xdy=e^{-y}\sec^2 y\,dy$ $\left[\textbf{Ans: } xe^y=\tan y+c\right]$

16. $(2x-10y^3)\dfrac{dy}{dx}+y=0$ $\left[\textbf{Ans: } x=2y^3+\dfrac{c}{y^2}\right]$

17. $(1+x^2)\dfrac{dy}{dx}=6x^2-2xy$ $\left[\textbf{Ans: } y(1+x^2)=2x^3+c\right]$

18. $x\dfrac{dy}{dx}+y\log y=xye^x$ $\left[\textbf{Ans: } x\log y=(x-1)e^x+c\right]$

19. $\sqrt{1-y^2}\,dx=(\sin^{-1}y-x)dy$ $\left[\textbf{Ans: } x=\sin^{-1}y-1+ce^{\sin^{-1}y}\right]$

20. $y\log y dx+(x-\log y)dy=0$ $\left[\textbf{Ans: } x=\dfrac{1}{2}\log y+\dfrac{c}{\log y}\right]$.

4.5.7 Bernoulli equation

Suppose the equation to be solved is of the form

$$\frac{dy}{dx}+Py = Qy^n \qquad \ldots 1$$

where P and Q are functions of x only . Dividing this equation y^n, we get

$$y^{-n}\frac{dy}{dx}+Py^{-n+1} = Q \qquad \ldots 2$$

An equation of the form (1) or (2) is called a ***"Bernoulli equation"***.
To solve this equation, we put

$$y^{-n+1} = v \qquad \dots 3$$

Then

$$(-n+1)y^{-n}\frac{dy}{dx} = \frac{dv}{dx} \qquad \dots 4$$

and equation (2) becomes

$$\frac{1}{1-n}\frac{dv}{dx}+Pv = Q$$

or

$$\frac{dv}{dx}+P(1-n)v = Q(1-n) \qquad \dots 5$$

This is a linear differential equation with v as the dependent variable. This equation may be solved by the procedure described in previous section i.e., *Linear Equation.*

WORKED EXAMPLES

Example 1 Solve $\frac{dy}{dx}+xy=xy^3$

Solution we first note that the given equation is of the form (1) (i.e., Bernoulli equation).Dividing by y^3, the equation becomes

$$\frac{1}{y^3}\frac{dy}{dx}+\frac{1}{y^2}x = x \qquad \dots i$$

Let us put

$$\frac{1}{y^2} = v \quad \text{So that } -\frac{2}{y^3}\frac{dy}{dx} = \frac{dv}{dx} \qquad \dots ii$$

Then equation (i) becomes

$$\left(-\frac{1}{2}\right)\frac{dv}{dx}+vx = x \quad \text{or} \quad \frac{dv}{dx}-2vx = -2x \qquad \dots iii$$

This is a linear equation in v and x with $P = -2x$ and $Q = -2x$
Therefore

$$\int P\,dx = \int -2x\,dx = -x^2$$

So that

$$\text{I.F} = e^{\int P\,dx} = e^{-x^2}$$

Hence, the general solution of (ii) is

$$v\,e^{\int P\,dy} = \int Q\,e^{\int P\,dy}\,dy+c$$

$$ve^{-x^2} = \int(-2x)e^{-x^2}\,dx + c$$
$$= \int e^t\,dt + c\text{, where } t = -x^2$$
$$= e^t + c$$

Where c is a constant. Recalling that $t = -x^2$, this becomes

$$ve^{-x^2} = e^{-x^2} + c$$

Using (ii), this becomes

$$\frac{1}{y^2}e^{-x^2} = e^{-x^2} + c \quad \text{or} \quad \frac{1}{y^2} = 1 + c\,e^{x^2}$$

This is the general solution of the given equation.

Example 2 Solve: $\dfrac{dy}{dx} + \dfrac{y}{x} = xy^2$

Solution The given equation is a Bernoulli equation. Dividing by y^2 the equation becomes

$$\frac{1}{y^2}\frac{dy}{dx} + \frac{1}{y}\cdot\frac{1}{x} = x \qquad \dots i$$

Let us put

$$\frac{1}{y} = v \quad \text{So that} \quad -\frac{1}{y^2}\frac{dy}{dx} = \frac{dv}{dx}$$

Then equation (i) becomes

$$\frac{dv}{dx} - \frac{1}{x}v = -x \qquad \dots ii$$

which is a linear equation in v and x with $P = -\dfrac{1}{x}$ and $Q = -x$

Therefore

$$\int P\,dx = -\int\frac{1}{x}\,dx = -\log x = \log\left(\frac{1}{x}\right)$$

So that $$\text{I.F} = e^{\int P\,dx} = e^{\log\left(\frac{1}{x}\right)} = \frac{1}{x}$$

Hence, the general solution of (ii) is

$$v\,e^{\int P\,dy} = \int Q\,e^{\int P\,dy}\,dy + c$$

$$v\frac{1}{x} = \int(-x)\frac{1}{x}\,dx + c = -\int 1dx + c = -x + c$$

Where c is a constant. substitute for v, this becomes

$$\frac{1}{xy} = -x + c$$

This is the general solution of the given equation.

Example 3 Solve $\frac{dy}{dx} + y\tan x = y^3 \sec x$

Solution Dividing by y^3, the equation becomes

$$\frac{1}{y^3}\frac{dy}{dx} + \frac{1}{y^2}\cdot\tan x = \sec x \qquad \ldots i$$

This is a Bernoulli equation
Let us put

$$\frac{1}{y^2} = v \quad \text{So that} \quad -\frac{2}{y^3}\frac{dy}{dx} = \frac{dv}{dx}$$

Then equation (i) becomes

$$\frac{dv}{dx} - 2(\tan x)v = -2\sec x \qquad \ldots ii$$

which is a linear equation in a and x with

$$P = -2\tan x \text{ and } Q = -2\sec x$$

Therefore

$$\int P\,dx = -\int 2\tan x dx = 2\log\cos x = \log\cos^2 x$$

So that $\quad \text{I.F} = e^{\int P\,dx} = e^{\log\cos^2 x} = \cos^2 x$

Hence, the general solution of (ii) is

$$e^{\int P\,dy} = \int Q\,e^{\int P\,dx}dy + c$$

$$v\cos^2 x = \int(-2\sec x)\cos^2 x\,dx + c$$

$$= -2\int\cos x\,dx + c$$

$$= 2\sin x + c$$

Where c is a constant. Substitute for $v = \frac{1}{y^2}$, this becomes

$$\frac{1}{y^2}\cos^2 x = 2\sin x + c$$

or $\quad \cos^2 x = y^2(2\sin x + c)$

This is the general solution of the given equation.

Example 4 Solve $x^3\frac{dy}{dx} - x^2 y = -y^4\cos x$

Solution Dividing by $-x^3 y^4$, the equation becomes

$$-\frac{1}{y^4}\frac{dy}{dx} + \frac{1}{y^3}\frac{1}{x} = \frac{1}{x^3}\cos x \qquad \ldots i$$

which is a Bernoulli equation.

Let us put

$$\frac{1}{y^3} = v \quad \text{So that} \quad -\frac{3}{y^4}\frac{dy}{dx} = \frac{dv}{dx}, \text{ the equation becomes}$$

$$\frac{dv}{dx} + \frac{3}{x}v = \frac{3}{x^3}\cos x \qquad \dots ii$$

which is a linear equation in v and x with

$$P = \frac{3}{x} \quad \text{and} \quad Q = \frac{3}{x^3}\cos x$$

Therefore

$$\int P\,dx = 3\int \frac{1}{x}dx = 3\log x$$

so that $\quad \text{I.F} = e^{\int P\,dx} = e^{3\log x} = x^3$

Hence, the general solution of (ii) is

$$ve^{\int P\,dy} = \int Qe^{\int P\,dy}dy + c$$

$$vx^3 = \int\left(\frac{3}{x^3}\cos x\right)x^3 dx + c$$

$$= 3\int \cos x\,dx + c = 3\sin x + c$$

Where c is a constant. Substitute for $v = \frac{1}{y^3}$, this becomes

$$\frac{1}{y^3}x^3 = 3\sin x + c$$

or $\quad x^3 = y^3(3\sin x + c)$

This is the general solution of the given equation.

Example 5 Solve $(3xy + y^2)dx - 3x^2dy = 0$

Solution The given equation may be rewritten as

$$\frac{dy}{dx} = \frac{3xy + y^2}{3x^2}$$

$$\frac{dy}{dx} = \frac{y}{x} + \frac{1}{3x^2}y^2$$

Which is a Bernoulli equation. Dividing by y^2 the equation becomes

$$\frac{1}{y^2}\frac{dy}{dx} - \frac{1}{x}\cdot\frac{1}{y} = \frac{1}{3x^2} \qquad \dots i$$

Let us put

$$\left(-\frac{1}{y}\right) = v, \quad \text{so that} \quad \frac{1}{y^2}\frac{dy}{dx} = \frac{dv}{dx}$$

Then equation (i) becomes

$$\frac{dv}{dx}+\frac{1}{x}v = \frac{1}{3x^2} \qquad \ldots ii$$

which is a linear equation in v and x with

$$P=\frac{1}{x} \quad \text{and} \quad Q = \frac{1}{3x^2}$$

Therefore

$$\int P\,dx = \int\frac{1}{x}dx=\log x$$

so that $\quad$ $\text{I.F} = e^{\int P\,dx} = e^{\log x} = x$

Hence, the general solution of (ii) is

$$ve^{\int P\,dy} = \int Q\,e^{\int P\,dy}+c$$

$$vx = \int\left(\frac{1}{3x^2}\right)x\,dx+c$$

$$= \frac{1}{3}\int\frac{1}{x}dx+c = \frac{1}{3}\log x+c$$

Where c is a constant. Substitute for $v=-\frac{1}{y}$, this becomes

$$-\frac{1}{y}x = \frac{1}{3}\log x+c$$

or

$$x = -y\left(\frac{1}{3}\log x+c\right)$$

This is the general solution of the given equation.

Example 6 Solve: $y\,(2xy + e^x)dx - xe^x dy = 0$

Solution The given equation may be rewritten as

$$\frac{dy}{dx} = \frac{y(2xy+e^x)}{xe^x}$$

$$\frac{dy}{dx} = \frac{2y^2}{e^x}+\frac{y}{x}$$

Which is a Bernoulli equation. Dividing by y^2, the equation becomes

$$\frac{1}{y^2}\frac{dy}{dx}-\frac{1}{x}\cdot\frac{1}{y} = \frac{2}{e^x} \qquad \ldots i$$

Let us put

$$\frac{1}{y}=v \quad \text{so that} \quad -\frac{1}{y^2}\frac{dy}{dx} = \frac{dv}{dx}$$

Then equation (i) becomes

$$\frac{dv}{dx}+\frac{1}{x}v = -2e^{-x} \qquad \ldots ii$$

which is a linear equation in with

$$P=\frac{1}{x} \quad \text{and} \quad Q = -2e^{-x}$$

Therefore

$$\int P\,dx = \int \frac{1}{x}dx = \log x$$

so that $\quad \text{I.F} = e^{\int P\,dx} = e^{\log x} = x$

Hence, the general solution of (ii) is

$$ve^{\int P\,dy} = \int Qe^{\int P\,dy}dy + c$$

$$vx = \int(-2e^{-x})x\,dx + c$$

$$= -2\int e^{-x}x\,dx + c$$

$$= 2(x+1)e^{-x} + c$$

Where c is a constant. Substitute for $v=\frac{1}{y}$, this becomes

$$\frac{1}{y}x = 2(x+1)e^{-x} + c$$

or $\quad x = y2(x+1)e^{-x} + c$

This is the general solution of the given equation.

Example 7 Solve $(x^2 + y^2 + x)dx + xy\,dy = 0$

The given equation may be rewritten as

$$\frac{dy}{dx} = \frac{x^2+y^2+x}{xy}$$

$$\frac{dy}{dx}+\frac{1}{x}y = -\frac{x+1}{y}$$

Which is a Bernoulli equation. $y^2 = v$, the equation becomes

so that $\quad 2y\frac{dy}{dx} = \frac{dv}{dx}$

The equation becomes

$$\frac{dv}{dx}+\frac{2}{x}v = -2(x+1)$$

which is a linear equation in *v* and *x* with

$$P=\frac{2}{x} \quad \text{and} \quad Q = -2(x+1)$$

Therefore

$$\int P\,dx = 2\int \frac{1}{x}dx = 2\log x$$

so that $\quad$ I.F. $= e^{\int P\,dx} = e^{2\log x} = x^2$

Hence, the general solution of (ii) is

$$v\,e^{\int P\,dy} = \int Q\,e^{\int P\,dy}dy + c$$

$$vx^2 = \int[-2(x+1)]x^2 dx + c$$

$$= -2\int(x+1)x^2 dx + c$$

$$= -2\int(x^3 + x^2)dx + c$$

$$= -2\left[\frac{x^4}{4} + \frac{x^3}{3}\right]dx + c$$

$$= -\left[\frac{x^4}{2} + \frac{2x^3}{3}\right]dx + c$$

Where c is a constant. Substitute for $v = y^2$, this becomes

$$y^2x^2 = -\left[\frac{x^4}{2} + \frac{2x^3}{3}\right]dx + c$$

or

$$y^2 = -\left[\frac{x^2}{2} + \frac{2x}{3}\right]dx + \frac{c}{x^2}$$

This is the general solution of the given equation.

Example 8 Solve: $(y\,\log x - 2)y\,dx = xdy$

Solution First we rewrite the given equation as

$$\frac{dy}{dx} = \frac{y}{x}(y\log x - 2) = \frac{-2y}{x} + \frac{1}{x}(\log x)y^2$$

or

$$\frac{1}{y^2}\frac{dy}{dx} + \frac{1}{y}\cdot\frac{2}{x} = \frac{1}{x}\log x$$

Which is a Bernoulli equation. Setting $\frac{1}{y} = v$, this equation becomes

$$\frac{dv}{dx} - \frac{2}{x}v = -\frac{1}{x}\log x$$

For this linear equation,

$$\text{I.F} = e^{\int(-2/x)}dx = e^{-2\log x} = \frac{1}{x^2}$$

Hence, its general solutions is

$$v\frac{1}{x^2} = \int\left(-\frac{1}{x}\log x\right)\left(\frac{1}{x^2}\right)dx + c = -\int\frac{1}{x^3}\log x + c$$

Where is a constant. Carrying out integration by parts, we obtain

$$\frac{v}{x^2} = -\left\{(\log x)\left(-\frac{1}{2x^2}\right) - \int\frac{1}{x}\left(-\frac{1}{2x^2}\right)dx\right\} + c$$

$$= \frac{1}{2x^2}\log x - \frac{1}{2}\int\frac{dx}{x^3} + c = \frac{1}{2x^2}\log x + \frac{1}{4x^2} + c$$

or $\quad v = \frac{1}{2}\log x + \frac{1}{4} + cx^2$

Substituting for v, this becomes

$$\frac{1}{y} = cx^2 + \frac{1}{2}\log x + \frac{1}{4}$$

This is the general solution of the given equation.

Example 9 Solve: $y - x\frac{dy}{dx} = y^2 + \frac{dy}{dx}$

Solution The given equation may be rewritten as

$$(1+x)\frac{dy}{dx} - y = -y^2$$

or $$-\frac{1}{y^2}\frac{dy}{dx} + \frac{1}{1+x}\cdot\frac{1}{y} = \frac{1}{1+x} \quad \ldots i$$

which is a Bernoulli equation Setting $v = \frac{1}{y}$ this equation becomes

$$\frac{dv}{dx} + \frac{1}{1+x}v = \frac{1}{1+x}$$

For this linear equation,

$$\int P\,dx = \int\frac{dx}{1+x} = \log(1+x)$$

So that $\quad \text{I.F} = e^{\int Pdx} = 1 + x$

Hence, the general solution of equation (i) is

$$v(1+x) = \int\frac{1}{1-x}(1+x)dx + c = x + c$$

Where c is a constant. Substituting for v, this becomes

$$\frac{1}{y} = \frac{c+x}{1+x}$$

This is the general solution of the given equation

Example10 Solve $\left(x^2y^3 + xy\right)\frac{dy}{dx} = 1$

The given equation may be rewritten as

$$\frac{dx}{dy} = x^2y^3 + xy \quad \text{or} \quad \frac{dx}{dy} - xy = y^3x^2$$

Dividing by x^2, we get

$$\frac{1}{x^2}\frac{dx}{dy} - \frac{1}{x}y = y^3 \qquad \ldots i$$

This is Bernoulli equation in which y is an independent variable and x is the dependent variable. Let us put

$$-\frac{1}{x} = v \qquad \ldots ii$$

So that $\frac{1}{x^2}\frac{dx}{dy} = \frac{dv}{dy}$. Then equation (i) becomes

$$\frac{dv}{dx} + yv = y^3 \qquad \ldots iii$$

This is a linear equation, in which y is an independent variable and v is the dependent variable, with $P = y,\ Q = y^3$. Therefore,

$$\int P\,dy = \frac{1}{2}y^2 \text{ and}$$

So that $\quad \text{I.F} = e^{\int P\,dy} = e^{y^2/2}$

Hence, the general solution of equation (iii) is given by

$$ve^{\int P\,dy} = \int y^3 e^{\int P\,dy}\,dy + c$$

i.e., $\quad ve^{y^2/2} = \int y^3 e^{y^2/2}dy + c = 2\int t\,e^t dt + c \quad \text{where } t = \frac{1}{2}y^3$

$$= 2(t-1)e^t + c = (y^2-2)e^{y^2/2} + c$$

Using (ii) this becomes

$$\frac{1}{x} = (2-y^2) - ce^{-(y^2/2)}$$

This is the general solution of the given equation

Example 11 Solve : $\frac{dz}{dx} + \frac{z}{x}\log z = \frac{z}{x^2}(\log z)^2$

Solution Dividing by z, the given equation becomes

$$\frac{1}{z}\frac{dz}{dx}+\frac{1}{x}\log z = \frac{1}{x^2}(\log z)^2 \qquad \ldots i$$

Setting

$$y = \log z \qquad \ldots ii$$

So that $\frac{dy}{dx} = \frac{1}{z}\frac{dz}{dx}$ equation (i) becomes

$$\frac{dy}{dx}+\frac{y}{x}=\frac{y^2}{x^2} \quad \text{or} \quad \frac{1}{y^2}\frac{dy}{dx}+\frac{1}{x}\frac{1}{y} = \frac{1}{x^2} \qquad \ldots iii$$

This is a Bernoulli equation Setting

$$\frac{1}{y} = v \qquad \ldots iv$$

We find that the general solution of this linear equation is

$$v\frac{1}{x} = \int\left(-\frac{1}{x^2}\right)\frac{1}{x}dx+c=\frac{1}{2}\cdot\frac{1}{x^2}+c$$

Using (iv) and (ii) this becomes

$$\frac{1}{\log z} = \frac{1}{2x}+cx$$

This is the general solution of the given equation

EXERCISES

Solve the following equations

1. $2\frac{dy}{dx}-\frac{y}{x}=\frac{y^2}{x^2}$ $\left[\textbf{Ans}: x = x+cy\sqrt{x}\right]$

2. $\frac{dy}{dx}-y=2xy^2e^{-x}$ $\left[\textbf{Ans}: e^x+y(x^2+x)=0\right]$

3. $\frac{dy}{dx}=x^3y^2+xy$ $\left[\textbf{Ans}: (-1/y)=x^2-2+ce^{-x/2}\right]$

4. $\frac{dy}{dx}+xy=x^3y^3$ $\left[\textbf{Ans}: 1/y^2=(1+x^2)+ce^{x^2}\right]$

5. $x\frac{dy}{dx}y=x^3y^6$ $\left[\textbf{Ans}: (/y^5)=(5/2)x^3+cx^5\right]$

6. $2x\frac{dy}{dx}=10x^3y^5+y$ $\left[\textbf{Ans}: x^2+(4x^5+c)y^4=0\right]$

7. $\dfrac{dy}{dx} = \dfrac{x^2 + y^2 + 1}{2xy}$ $\left[\textbf{Ans}: y^2 = x^2 + cx - 1\right]$

8. $\dfrac{dy}{dx} - y\tan x = y^2 \sec x$ [**Ans :** $\sec x = -y(c + \tan x)$]

9. $x\dfrac{dy}{dx} + y = y^2 \log x$ [**Ans :** $y(1 + cx + \log x) = 1$]

10. $x\dfrac{dy}{dx} + y = y^2 x^3 \cos x$ [**Ans :** $xy(x\sin x + \cos x + c) + 1 = 0$]

11. $xy - \dfrac{dy}{dx} = y^3 e^{-x^2}$ [**Ans :** $(1/y^2) = (2x + c)e^{-x^2}$]

12. $x^3 \dfrac{dy}{dx} + y^4 \cos x = x^2 y$ [**Ans :** $x^3 = (c + 3\sin x)y^3$]

13. $3y^2 \dfrac{dy}{dx} + 2xy^3 = 4xe^{-x^2}$ [**Ans :** $y^3 = (2x^2 + c)e^{-x^2}$]

14. $\dfrac{dy}{dx} - y\tan x = \dfrac{\sin x}{y^2 \sec^2 x}$ [**Ans :** $y^3 \cos^3 x = c - (1/2)\cos^6 x$]

15. $\dfrac{dy}{dx} + xy = y^2 e^{x^{2/2}} \log x$ [**Ans :** $y(x - x\log x + c) = e^{-x^2/2}$]

16. $\dfrac{dy}{dx} + x(x + y) = x^3(x + y)^3 - 1$ [**Ans :** $(x + y)^{-2} = (x^2 + 1) + ce^{x^2}$]

17. $x(x - y)dy + y^2\, dx = 0$ [**Ans :** $y/x = \log y + c$]

18. $\cos x\, dy = y\,(\sin x - y)dx$ [**Ans :** $\sec x = y(c + \tan x)$]

19. $y(2xy + e^x)dx - e^x\, dy = 0$ [**Ans :** $y(c + x^2) + e^x = 0$]

20. $y(x^2 y + e^x)\, dx = e^x\, dy$ [**Ans :** $x^3 y + 3e^x = cy$]

4.6 Linear Differential Equations of Second and Higher Order with Constant Coefficients

A differential equation of the form

$$\frac{d^n y}{dx^n} + a_1 \frac{d^{n-1} y}{dx^{n-1}} + a_2 \frac{d^{n-2} y}{dn^{n-2}} + \ldots + a_n y = X \qquad \ldots.\ 1$$

where X is a function of x and a_1, a_2, ... a_n are constants is called a linear differential equation of nth order with constant coefficients. Since the highest order derivative appearing in (1) is n, it is called a differential equation of nth order and it is called linear.

Using the familiar notation of differential operators:

$$D = \frac{d}{dx},\ D^2 = \frac{d^2}{dx^2},\ D^3 = \frac{d^3}{dx^3},\ D^4 = \frac{d^4}{dx^4}, \ldots D^n = \frac{d^n}{dx^n}$$

Then (1) can be written in the form

$$\{D^n + a_1 D^{n-1} + a_2 D^{n-2} + \ldots + a_n\}y = X \qquad \ldots 2$$

i.e., $\quad f(D)y = X$

where $\quad f(D) = D^n + a_1 D^{n-1} + a_2 D^{n-2} + \ldots + a_n$

Here $f(D)$ is a polynomial of degree n in D

If $x = 0$, the equation $f(D)y = 0$ is called a homogeneous equation.

If $x \neq 0$ the equation (2) is called a non- homogeneous equation

4.7 Solution of a Homogeneous Second Order Linear Differential Equations

We consider the homogeneous equation

$$\frac{d^2y}{dx^2} + p\frac{dy}{dx} + qy = 0 \qquad \ldots 1$$

where p and q are constants.

$$(D^2 + pD + q)y = 0 \qquad \ldots 2$$

The Auxiliary equations (A.E) put $D = m$

$$m^2 + pm + q = 0 \qquad \ldots 3$$

Equation (3) is called auxiliary equation (A.E) or characteristic equation of the D.E [equation (3)] being quadratic in m, will have two roots in general. There are three cases.

Case (i) : Roots are real and distinct

The roots are real and distinct, say m_1 and m_2 i.e., $m_1 \neq m_2$

Hence, the general solution of equation (1) is

$$y = C_1 e^{m_1 x} + C_2 e^{m_2 x}$$

where C_1 and C_2 are arbitrary constant.

Case (ii) : Roots are equal.

The roots are equal, say m_1 and m_2 i.e., $m_1 = m_2 = m$

Hence, the general solution of equation (1) is

$$y = (C_1 + C_2 x)e^{mx}$$

where C_1 and C_2 are arbitrary constant.

Case (iii) : Roots are complex.

The roots are complex , say $\alpha \pm \beta$

Hence , the general solution of equation (1) is

$$y = e^{\alpha x}(C_1 \cos\beta x + C_2 \sin\beta x)$$

where C_1 and C_2 are arbitrary constant.

Note: Complimentary Function (C.F) which itself is the general solution of the D.E

WORKED EXAMPLES

1. Solve: $\dfrac{d^2y}{dx^2}-5\dfrac{dy}{dx}+6y=0$

Solution Given equation is $(D^2-5D+6)y=0$

A.E is $m^2-5m+6 = 0$

i.e., $(m-2)(m-3) = 0$

i.e., $m = 2, 3$

$\therefore$ $m_1=2,\ m_2 = 3$

Therefore the roots are real and distinct.
Hence, the general solution of equation is

$$y = C_1e^{2x}+C_2e^{3x}$$

2. Solve: $\dfrac{d^3y}{dx^3}-\dfrac{d^2y}{dx^2}-4\dfrac{dy}{dx}+4y=0$

Solution Given equation is $(D^3-D^2-4D+4)y=0$

A.E is $m^3-m^2-4m+4 = 0$

$m^2(m-1)-4(m-1) = 0$

$(m-1)(m^2-4) = 0$

$m=1,\quad m = \pm 2$

$m_2=1, m_2=2, m_3 = -2$

Therefore the roots are real and distinct.
Hence, the general solution of equation is

$$y = C_1e^{x}+C_2e^{2x}+C_3e^{-2x}$$

3. Solve: $\dfrac{d^2y}{dx^2}-\dfrac{dy}{dx}-6y=0$

Solution The D.E. can be written as $(D^2-D-6)y=0$

A.E is $m^2-m-6 = 0$

$(m-3)(m+2) = 0$

$m = 3, -2$

$m_1=3,\ m_2 = -2$

Therefore the roots are real and distinct.
Hence, the general solution of equation is

$$y = C_1e^{3x}+C_2e^{-2x}$$

4. Solve: $\dfrac{d^2y}{dx^2}+8\dfrac{dy}{dx}+16y=0$

Solution The D.E. can be written as $(D^2 + 8D + 16)y = 0$

$$\begin{aligned} \text{A.E is } m^2 + 8m + 16 &= 0 \\ (m + 4)^2 &= 0 \\ m &= -4, -4 \\ m_1 = m_2 &= -4 \end{aligned}$$

Therefore the roots are Equal

Hence, the general solution of equation is

$$y = (C_1 + C_2 x)e^{-4x}$$

5. Solve: $\dfrac{d^2y}{dx^2} + w^2 y = 0$

Solution The D.E. can be written as $(D^2 + w^2)y = 0$

$$\begin{aligned} \text{A.E is } m^2 + w^2 &= 0 \\ m^2 &= -w^2 \\ m^2 &= w^2 i^2 \\ m_1 &= \pm wi \end{aligned}$$

Therefore the roots are Complex and this is the form $\alpha \pm \beta$, where $\alpha = 0, \beta = w$

Hence, the general solution of equation is

$$y = e^{0x}(C_1 \cos wx + C_2 \sin wx)$$

$$\therefore \quad y = C_1 \cos wx + C_2 \sin wx$$

6. Solve: $\dfrac{d^2y}{dx^2} + 4\dfrac{dy}{dx} + 13y = 0$

Solution : The D.E. can be written as $(D^2 + 4D + 13)y = 0$

$$\text{A.E. is } m^2 + 4m + 13 = 0$$

$$m = \frac{-4 \pm \sqrt{16 - 52}}{2} \qquad \left[\text{By using the formula } \frac{-b \pm \sqrt{b^2 - 4ac}}{2a}\right]$$

$$m = -2 \pm 3i$$

Therefore the roots are Complex and this is the form $\alpha \pm \beta$, where $\alpha = -2, \beta = 3$

Hence, the general solution of equation is

$$y = e^{-2x}(C_1 \cos 3x + C_2 \sin 3x)$$

7. Solve: $\dfrac{d^2y}{dx^2} - \dfrac{dy}{dx} = 0$ given $y'(0) = 0,\ y(0) = 1$

Solution The D.E. can be written as $(D^2 - D)y = 0$

$$\begin{aligned} \text{A.E. is } m^2 - m &= 0 \\ m(m - 1) &= 0 \\ m &= 0, 1 \end{aligned}$$

Therefore the roots are real and distinct.

Hence, the general solution of equation is

$$y = C_1 e^{0x} + C_2 e^{1x}$$

$\therefore$ $y = C_1 + C_2 e^x$

i.e., $x = 0,\ y = 0$ (Given)

When $y(0) = 0$

$$0 = C_1 + C_2 \qquad \ldots 1$$

$\Rightarrow$ $y'(x) = C_2 e^x$

Given, when $x = 0,\ y' = 1$

$$y'(0) = 1$$

$\Rightarrow$ $1 = C_2 e^0$

$$C_2 = 1 \qquad \ldots 2$$

From (1) and (2) $\Rightarrow C_1 = -1$

$\therefore$ The general solution of equation is $y = e^x - 1$

8. Solve: $(4D^4 - 4D^3 - 23D^2 + 12D + 36)y = 0$

Solution Given equation is $(4D^4 - 4D^3 - 23D^2 + 12D + 36)y = 0$

A.E. is $(4m^4 - 4m^3 - 23m^2 + 12m + 36 = 0$

If $m = 2,\ 64 - 32 - 92 + 24 + 36 = 0$

$\Rightarrow$ $m = 2$ is root of inspection

By synthetic division,

2	4	–4	–23	12	36
	0	8	+8	–30	–36
	4	+4	–15	–18	0

A.E is $4m^2 + 4m - 15m - 18 = 0$

If $m = 2,\ 32 + 16 - 30 - 18 = 0$

$\Rightarrow$ Again $m = 2$ is a root

By synthetic division,

2	4	4	–15	–18
	0	8	24	18
	4	12	9	0

$\therefore$ $4m^2 + 12m + 9 = 0$

A.E is $(2m + 3)^2 = 0$

$$m_1 = m_2 = m = -\frac{3}{2}$$

Therefore the roots of the A.E are $2, 2, -\frac{3}{2}, -\frac{3}{2}$.

Hence, the general solution of equation is

$$y = (C_1 + C_2 x)e^{2x}(C_3 + C_4 x)e^{-\frac{3}{2}x}$$

9. Solve: $(D^5 - D^4 - D + 1)y = 0$

Solution Given equation is $(D^5 - D^4 - D + 1)y = 0$

$$\begin{aligned}\text{A.E. is } m^5 - m^4 - m + 1 &= 0\\ m^4(m-1) - 1\,(m-1) &= 0\\ (m^4 - 1)\,(m-1) &= 0\\ (m^2-1)\,(m^2+1)\,(m-1) &= 0\\ (m-1)\,(m+1)\,(m^2+1)\,(m-1) &= 0\\ m = 1, 1, -1, \pm i\end{aligned}$$

Therefore the roots of the A.E are 1, 1, – 1, ± i

Hence, the general solution of equation is

$$y = (C_1 + C_2x)e^x + C_3e^{-x} + (C_4\cos x + c_3\sin x)$$

EXERCISES

1. $\dfrac{d^2y}{dx^2} - 2\dfrac{dy}{dx} + 3y = 0$ $\left[\textbf{Ans: } y = C_1e^{3x} + C_2e^{-x}\right]$

2. $6\dfrac{d^2y}{dx^2} - \dfrac{dy}{dx} - y = 0$ $\left[\textbf{Ans: } y = C_1e^{\frac{1}{2}x} + C_2e^{-\frac{1}{3}x}\right]$

3. $2\dfrac{d^2y}{dx^2} - \dfrac{dy}{dx} - 6y = 0$ $\left[\textbf{Ans: } y = C_1e^{2x} + C_2e^{-\frac{3}{2}x}\right]$

4. $\dfrac{d^2y}{dx^2} + 4\dfrac{dy}{dx} + 4y = 0$ $\left[\textbf{Ans: } y = (C_1 + C_2)e^{-2x}\right]$

5. $9\dfrac{d^2y}{dx^2} - 6\dfrac{dy}{dx} + 4y = 0$ $\left[\textbf{Ans: } y = (C_1 + C_2x)e^{\frac{1}{3}x}\right]$

6. $\dfrac{d^2y}{dx^2} + 9y = 0$ $\left[\textbf{Ans: } y = C_1\cos 3x + C_2\sin 3x\right]$

7. $\dfrac{d^2y}{dx^2} - 2\dfrac{dy}{dx} + 2y = 0$ $\left[\textbf{Ans: } y = e^x(C_1\cos x + C_2\sin x)\right]$

8. $\dfrac{d^2y}{dx^2} + \dfrac{dy}{dx} + y = 0$ $\left[\textbf{Ans: } y = e^{-\frac{1}{2}x}\left(C_1\cos\dfrac{\sqrt{3}}{2}x + C_2\sin\dfrac{\sqrt{3}}{2}x\right)\right]$

9. $\dfrac{d^2y}{dx^2} + 4\dfrac{dy}{dx} + 5y = 0$ $\left[\textbf{Ans: } y = e^{-2x}(C_1\cos x + C_2\sin x)\right]$

10. $\dfrac{d^3y}{dx^3} - 8y = 0$ $\left[\textbf{Ans: } y = C_1e^{2x} + e^{-x}\left(C_1\cos\sqrt{3}x + C_2\sin\sqrt{3}x\right)\right]$

11. $\dfrac{d^3y}{dx^3}+6\dfrac{d^2y}{dx^2}+11\dfrac{dy}{dx}+6y=0$ $\left[\textbf{Ans: } y=C_1e^{-x}+C_2e^{-2x}+C_3e^{-3x}\right]$

12. $\dfrac{d^3y}{dx^3}-4\dfrac{d^2y}{dx^2}+5\dfrac{dy}{dx}-2y=0$ $\left[\textbf{Ans: } y=(C_1+C_2x)e^{x}+C_3e^{2x}\right]$

13. $\dfrac{d^3y}{dx^3}+6\dfrac{d^2y}{dx^2}+12\dfrac{dy}{dx}+8y=0$ $\left[\textbf{Ans: } y=(C_1+C_2x+C_3x^2)e^{-2x}\right]$

14. $\dfrac{d^4y}{dx^4}-2\dfrac{d^3y}{dx^3}+5\dfrac{d^2y}{dx^2}-8\dfrac{dy}{dx}+4y=0$

$\left[\textbf{Ans: } y=(C_1+C_2x)e^{x}+(C_3\cos 2x+C_4\sin 2x)\right]$

15. $\dfrac{d^4y}{dx^4}-4\dfrac{d^3y}{dx^3}+8\dfrac{d^2y}{dx^2}-8\dfrac{dy}{dx}+4y=0$

$\left[\textbf{Ans: } y=e^{x}[(C_1+C_2x)\cos x+(C_3+C_4x)\sin x]\right]$

16. $\dfrac{d^4y}{dx^4}-\dfrac{d^3y}{dx^3}-9\dfrac{d^2y}{dx^2}-11\dfrac{dy}{dx}-4y=0$

$\left[\textbf{Ans: } y=(C_1+C_2x+C_3x^2)e^{-x}+C_4e^{4x}\right]$

4.8 Inverse Differential Operator and Particular Integral

Consider a differential equation

$$f(D)y = x \qquad \dots 1$$

Define $\dfrac{1}{f(D)}$ such that

$$f(D)\left\{\frac{1}{f(D)}\right\}x = x \qquad \dots 2$$

Here $f(D)$ is called the inverse diffrential operator. Hence from Eqn. (1), we obtain

$$y = \frac{1}{f(D)}x \qquad \dots 3$$

Since this satisfies the Eqn. (1) hence the particular integral of Eqn. (1) is given by Eqn. (3)

Thus, particular Integral (P.I.) $=\dfrac{1}{f(D)}x$

The inverse differential operator $\frac{1}{f(D)}$ is linear.

i.e., $\frac{1}{f(D)}\{ax_1 + bx_2\} = a\frac{1}{f(D)}x_1 + b\frac{1}{f(D)}x_2$

where a, b are constants and x_1 and x_2 are some funtions of x.

4.8.1 Special forms of X

Type 1: P.I. of the form $\frac{e^{ax}}{f(D)}$

We have the equation $f(D) = e^{ax}$

$$
\begin{aligned}
\text{Let} \qquad f(D) &= D^2 + a_1 D + a_2 \\
\text{We have} \qquad D(e^{ax}) &= ae^{ax}, D^2(e^{ax}) = a^2 e^{ax} \text{ and so on.} \\
\therefore \qquad f(D)(e^{ax}) &= \left(D^2 + a_1 D + a_2\right)e^{ax} \\
&= a^2 e^{ax} + a_1 \cdot ae^{ax} + a_2 e^{ax} \\
&= \left(a^2 + a_1 \cdot a + a_2\right)e^{ax} = f(a)e^{ax} \\
\text{Thus} \qquad f(b)e^{ax} &= f(a)e^{ax}
\end{aligned}
$$

Operating with $\frac{1}{f(D)}$ on both sides

$$
\begin{aligned}
\text{we get,} \qquad e^{ax} &= f(a)\cdot\frac{1}{f(D)}\cdot e^{ax} \\
\text{P.I.} &= \frac{1}{f(D)}e^{ax} = \frac{e^{ax}}{f(D)}
\end{aligned}
$$

In particular if $f(D) = D - a$, then using the general formula.

$$
\text{We get,} \qquad \frac{1}{D-a}e^{ax} = \frac{e^{ax}}{(D-a)\phi(D)} = \frac{1}{D-a}\cdot\frac{e^{ax}}{\phi(a)}
$$

$$
\begin{aligned}
\text{i.e.,} \qquad \frac{e^{ax}}{f(D)} &= \frac{1}{\phi(a)}e^{ax}\int 1\cdot dx = \frac{1}{\phi(a)}\cdot xe^{ax} \\
\therefore \qquad f'(a) &= 0 + \phi(a) \\
f'(a) &= \phi(a)
\end{aligned}
$$

or

Thus, Eqn. (1) becomes

$$
\frac{e^{ax}}{f(D)} = x\cdot\frac{e^{ax}}{f'(D)}
$$

$$
\begin{aligned}
\text{where} \qquad f(a) &= 0 \\
\text{and} \qquad f'(a) &\neq 0
\end{aligned}
$$

This result can be extended fruther also if

$$f'(a) = 0, \frac{e^{ax}}{f(D)}$$

$$= x^2 \cdot \frac{e^{ax}}{f''(a)} \text{ and so on.}$$

Type 2: P.I. of the form $\frac{\sin ax}{f(D)}, \frac{\cos ax}{f(D)}$

We have

$$D(\sin ax) = a\cos ax$$

$$D^2(\sin ax) = -a^2 \sin ax$$

$$D^3(\sin ax) = -a^3 \cos ax$$

$$D^4(\sin ax) = a^4 \sin ax$$

$$= (-a^2)^2 \sin ax \text{ and so on.}$$

Therefore, if $f(D^2)$ is a rational integral function of D^2 then $f(D^2)\sin ax = f(-a^2)\sin ax$

Hence $\frac{1}{f(D^2)}\{f(D^2)\sin ax\} = \frac{1}{f(D^2)} f(-a^2)\sin ax$

i.e.,

$$\sin ax = f(-a^2)\frac{1}{f(D^2)}\sin ax$$

i.e.,

$$\frac{1}{f(D^2)}\sin ax = \frac{\sin ax}{f(-a^2)}$$

Provided $f(-a^2) \neq 0$ ... 1

Similarly, we can prove that

$$\frac{1}{f(D^2)}\cos ax = \frac{\cos ax}{f(-a^2)}$$

If $f(-a^2) \neq 0$ ... 2

In general,

$$\frac{1}{f(D^2)}\cos ax = \frac{\cos ax}{f(-a^2)}$$

If $f(-a^2) \neq 0$

$$\frac{1}{f(D^2)}\sin(ax+b) = \frac{1}{f(-a^2)}\sin(ax+b)$$

and

$$\frac{1}{f(D^2)}\cos(ax+b) = \frac{1}{f(-a^2)}\cos(ax+b)$$

These formula can be easily remambered as follows :

$$\frac{1}{D^2+a^2}\sin ax = \frac{x}{2}\int \sin axdx = \frac{-x}{2a}\cos ax$$

$$\frac{1}{D^2+a^2}\cos ax = \frac{x}{2}\int \cos axdx = \frac{x}{2a}\sin ax$$

Type 3: P.I. of the form $\frac{\phi(x)}{f(D)}$ where $\phi(x)$ is a polynomial in x, we seeking the polynomial Eqn. as the particular soultion of

$$f(D)y = \phi(x)$$

where $\phi(x) = a_0x^n + a_1x^{n-1} + \ldots a_{n-1}x + a_n$

Hence P.I. is found by divisor. By writing $\phi(x)$ in descending powers of x and $f(D)$ in ascending powers of D. The division get completed without any remainder. The quotient so obtained in the process of division will be particular integral.

Type 4: P.I. of the form $\frac{e^{ax}V}{f(D)}$ where V is a function of x.

We shall prove that $\frac{1}{f(D)}e^{ax}V = e^{ax}\frac{1}{f(D+a)}V$.

$$\begin{aligned}
\text{Consider} \quad D(e^{ax}V) &= e^{ax}DV + Vae^{ax} \\
&= e^{ax}(D+a)V \\
\text{and} \quad D^2(e^{ax}V) &= e^{ax}D^2V + ae^{ax}DV + a^2e^{ax}V + ae^{ax}DV \\
&= e^{ax}(D^2V + 2aDV + a^2V) \\
&= e^{ax}(D+a^2)V \\
\text{Similarly,} \quad D^3(e^{ax}V) &= e^{ax}(D+a)^3V \text{ and so on.} \\
\therefore \quad f(D)e^{ax}V &= e^{ax}f(D+a)V \qquad \ldots 1 \\
\text{Let} \quad f(D+a)V &= U, \\
\text{so that} \quad V &= \frac{1}{f(D+a)}U
\end{aligned}$$

Hence (1) reduces to

$$f(D)e^{ax}\frac{1}{f(D+a)}U = e^{ax}U$$

Operating both sides by $\frac{1}{f(D)}$ we get,

$$e^{ax}\frac{1}{f(D+a)}U = \frac{1}{f(D)}e^{ax}U$$

i.e., $$\frac{1}{f(D)} e^{ax} U = e^{ax} \frac{1}{f(D+a)} U$$

Replacing U by V, we get the required result.

Type 5: P.I. of the form $\frac{xV}{f(D)}, \frac{x^n V}{f(D)}$ whee V is a function of x.

By Leibniz's theorem, we have

$$D^n (xV) = xD^n V + n \cdot 1 D^{n-1} \cdot V$$

$$= xD^n V + \left\{ \frac{d}{dD} D^n \right\} V$$

$$\therefore \quad f(D)xV = xf(D)V + f'(D)V \qquad \ldots 1$$

Eqn. (1) reduces to

$$\frac{xV}{f(D)} = \left[x \frac{f'(D)}{f(D)} \right] \frac{V}{f(D)} \qquad \ldots 2$$

This is formula for finding the particular integral of the functions of the xV. By repeated application of this formula, we can find P.I. as $x^2 V, x^3 V \ldots.$

WORKED EXAMPLES

Type 1

1. Solve: $\frac{d^2 y}{dx^2} - 5\frac{dy}{dx} + 6y = e^{5x}$

Solution We have $(D^2 - 5D + 6)y = e^{5x}$

A.E. is $m^2 - 5m + 6 = 0$

i.e. $(m-2)(m-3) = 0$

$m = 2, 3$

Hence the complementary function is

$$\text{C.F} = C_1 e^{2x} + C_2 e^{3x}$$

Particular integral (P. I) is

$$\text{P.I} = \frac{1}{D^2 - 5D + 6} e^{5x} \qquad (D \to 5)$$

$$= \frac{1}{5^2 - 5 \times 5 + 6} e^{5x} = \frac{e^{5x}}{6}$$

$\therefore$ The general solution is given by

$$y = \text{C.F} + \text{P.I}$$

$$= C_1 e^{2x} + C_2 e^{3x} + \frac{e^{5x}}{6}$$

2. Solve: $\frac{d^2y}{dx^2} - 3\frac{dy}{dx} + 2y = 10e^{3x}$

Solution We have

$$(D^2 - 3D + 2)y = 10e^{3x}$$

$$\text{A.E. is } m^2 - 3m + 2 = 0$$

$$\text{i.e., } (m-2)(m-1) = 0$$

$$m = 2, 1$$

$$\text{C.F} = C_1e^{2x} + C_2e^{1x}$$

$$\text{P.I} = \frac{1}{D^2 - 3D + 2}10e^{3x} \quad (D \to 3)$$

$$= \frac{1}{3^2 - 3\times 3 + 2}10e^{3x} = \frac{10e^{3x}}{2}$$

$\therefore$ The general solution is given by

$$y = \text{C.F} + \text{P.I.}$$

$$= C_1e^{2x} + C_2e^{1x} + \frac{10e^{3x}}{2}$$

3. Solve: $\frac{d^2y}{dx^2} - 4\frac{dy}{dx} + 4y = e^{2x}$

Solution Given equation is $(D^2 - 4D + 4)y = e^{2x}$

$$\text{A.E. is } m^2 - 4m + 4 = 0$$

i.e.,

$$(m-2)(m-2) = 0$$

$$m = 2, 2$$

$$\text{C.F} = (C_1 + C_2)e^{2x}$$

$$\text{P.I} = \frac{1}{D^2 - 4D + 4}e^{2x} \quad (D = 2)$$

$$= \frac{1}{2^2 - 4(2) + 4}e^{2x} \quad (Dr = 0)$$

Differentiate the denominator and multiply 'x'

$$= x \cdot \frac{1}{2D - 4}e^{2x} \quad (D = 2)$$

$$= x \cdot \frac{1}{2(2) - 4}e^{2x} \quad (D = 0)$$

Again differentiate denominator and multiply 'x'

$$= x^2 \frac{1}{2}e^{2x}$$

$$\text{P.I} = x^2\frac{1}{2}e^{2x}$$

$$y = C.F + P.I = (C_1 + C_2)e^{2x} + \frac{x^2e^{2x}}{2}$$

4. Solve: $\dfrac{d^4x}{dt^4} + 4x = \cosh t$

Solution Given equation is $(D^4 + 4)x = \cos ht$

$$\text{A.E. is } m^4 + 4 = 0$$

i.e., $(m^2 + 2)^2 - 4m^2 = 0$

or $[(m^2 + 2) - 2m]\,[(m^2 + 2) + 2m] = 0$

$$m^2 - 2m + 2 = 0 \;;\; m^2 + 2m + 2 = 0$$

$$m = \frac{-b \pm \sqrt{b^2 - 4ac}}{2a}$$

$$m = \frac{2 \pm \sqrt{4-8}}{2}; \quad m = \frac{-2 \pm \sqrt{4-8}}{2}$$

$$m = \frac{2 \pm 2i}{2}; \quad m = \frac{-2 \pm 2i}{2}$$

$$m = 1 \pm i; \quad m = -1 \pm i$$

$$\text{C.F} = e^t(C_1 \cos t + C_2 \sin t) + e^{-t}(C_3 \cos t + C_4 \sin t)$$

$$\text{P.I} = \frac{1}{D^4 + 4}\cosh t$$

$$\text{Where } \cosh t = \frac{e^t + e^{-t}}{2}$$

$$P.I = \frac{1}{2}\left[\frac{e^t + e^{-t}}{D^4 + 4}\right]$$

$$= \frac{1}{2}\left[\frac{e^t}{D^4 + 4}\right] + \frac{1}{2}\left[\frac{e^{-t}}{D^4 + 4}\right] \qquad (D \to 1)\,(D \to -1)$$

$$= \frac{1}{2}\left[\frac{e^t}{5}\right] + \frac{1}{2}\left[\frac{e^{-t}}{5}\right]$$

$$= \frac{1}{5}\cdot\frac{1}{2}[e^t + e^{-t}] = \frac{1}{5}\cosh t$$

$$x = \text{C.F} + \text{P.I}$$

$$x = e^t(C_1 \cos t + C_2 \sin t) + e^{-t}(C_3 \cos t + C_4 \sin t) + \frac{1}{5}\cosh t$$

5. Solve: $\frac{d^3y}{dx^3} - y = (e^x + 1)^2$

Solution Given equation can be written as

$$(D^3 - 1)y = e^{2x} + 2e^x + 1$$

$$\text{A.E. is } m^3 - 1 = 0$$

$$\text{i.e., } (m-1)(m^2 + m + 1) = 0$$

Hence $m - 1 = 0$, and $m^2 + m + 1 = 0$

$$m = \frac{-b \pm \sqrt{b^2 - 4ac}}{2a}$$

$$m = \frac{-1 \pm \sqrt{1-4}}{2}$$

$$m = \frac{-1 \pm \sqrt{3}i}{2} \qquad \therefore\ m = 1\ ;\ m = \frac{-1 \pm \sqrt{3}i}{2}$$

$$C.F. = C_1 e^x + \left(C_2 \cos \frac{\sqrt{3}}{2} x + C_2 \sin \frac{\sqrt{3}}{2} x \right) e^{-\frac{1}{2}x}$$

$$P.I. = \frac{e^{2x}}{D^3 - 1} + \frac{2e^x}{D^3 - 1} + \frac{1}{D^3 - 1}$$

$$= P.I._1 + P.I._2 + P.I._3$$

$$P.I._1 = \frac{e^{2x}}{D^3 - 1} \quad (D \to 2)$$

$$= \frac{e^{2x}}{2^3 - 1} = \frac{e^{2x}}{7}$$

$$P.I._2 = \frac{1}{D^3 - 1} 2e^x \quad (D \to 1)$$

$$= \frac{1}{1^3 - 1} 2e^x \quad (Dr = 0)$$

Differentiate the *Dr* and multiply *x*

$$= \frac{2xe^x}{3D^2} \quad (D \to 1)$$

$$P.I._2 = x \cdot \frac{2e^x}{3} \quad (D \to 1)$$

$$P.I._3 = \frac{1}{D^3-1}e^0 \quad (D \to 0) = -1$$

$$P.I. = \frac{e^{2x}}{7} + \frac{2xe^x}{3} - 1$$

$\therefore$ The general solution is

$$y = C.F. + P.I.$$

$$= C_1e^x + e^{-\frac{1}{2}x}\left[C_2 \cos\frac{\sqrt{3}}{2}x + C_3 \sin\frac{\sqrt{3}}{2}x\right] + \frac{e^{2x}}{7} + \frac{2}{3}xe^x - 1$$

6. Solve: $\left(D^3 + 2D^2 - D - 2\right)y = 2\cosh x$

Solution Given equation is

$$\left(D^3 + 2D^2 - D - 2\right)y = 2\cosh x$$

$$\text{A.E. is } m^3 + 2m^2 - m - 2 = 0$$

$$m^2(m+2) - 1(m+2) = 0$$

$$(m+2)(m^2-1) = 0$$

$$(m+2)(m+1)(m-1) = 0$$

$$m = -2, -1, 1$$

$$C.F. = C_1e^{-2x} + C_2e^{-x} + C_3e^x$$

$$P.I. = \frac{1}{D^3 + 2D^2 - D - 2}2\cosh x$$

$$\left(\text{where} \cosh x = \frac{e^x + e^{-x}}{2}\right)$$

$$= \frac{2}{D^3 + 2D^2 - D - 2}\left(\frac{e^x + e^{-x}}{2}\right)$$

$$= \frac{e^x}{D^3 + 2D^2 - D - 2} + \frac{e^{-x}}{D^3 + 2D^2 - D - 2}$$

$$= P.I_1 + P.I._2$$

$$P.I_1 = \frac{e^x}{1^3 + 2(1)^2 - 1 - 2} \quad D \to 1$$

$$= \frac{e^x}{1^3 + 2(1)^2 - 1 - 2} \quad Dr = 0$$

$$= \frac{xe^x}{3D^2+4D-1} \quad D \to 1$$

$$= \frac{xe^x}{3(1)^2+4(1)-1} = \frac{xe^x}{6}$$

$$P.I._2 = \frac{e^{-x}}{D^3+2D^2-D-2} \quad D \to -1$$

$$= \frac{e^{-x}}{(-1)^3+2(-1)^2-(-1)-2} \quad Dr = 0$$

$$= \frac{xe^{-x}}{3D^2+4D-1} = \frac{xe^{-x}}{3(1)^2+4(-1)-1} \qquad (D \to -1)$$

$$= -\frac{xe^{-x}}{2}$$

$$P.I. = \frac{xe^x}{6} - \frac{xe^{-x}}{2}$$

$$y = C.F. + P.I.$$

$$y = C_1e^{-2x} + C_2e^{-x} + C_3e^x + \frac{xe^x}{6} - \frac{xe^{-x}}{2}$$

Type 2:

1. Solve: $(D^2+9)y = \cos 4x$

Solution: Given equation is $(D^2+9)y = \cos 4x$

A.E is $m^2 + 9 = 0$

$$m^2 = -9$$

$$m = \pm 3i$$

$$\text{C.F.} = C_1 \cos 3x + C_2 \sin 3x$$

$$P.I. = \frac{1}{D^2+9}\cos 4x \quad D^2 = -4^2 = -16$$

$$= \frac{1}{-16+9}\cos 4x$$

$$= -\frac{1}{7}\cos 4x$$

$$y = C.F. + P.I.$$

$$y = C_1 \cos 3x + C_2 \sin 3x - \frac{1}{7}\cos 4x$$

2. Solve: $(D^2 + D + 1)y = \sin 2x$

Solution Given equation is $(D^2 + D + 1)y = \sin 2x$

$$\text{A.E. is } m^2 + m + 1 = 0$$

$$m = \frac{-1 \pm \sqrt{1-4}}{2}$$

$$= \frac{-1 \pm \sqrt{3}i}{2}$$

$$\text{Hence C.F.} = e^{-\frac{x}{2}}\left[C_1 \cos\frac{\sqrt{3}}{2}x + C_2 \sin\frac{\sqrt{3}}{2}x\right]$$

$$P.I. = \frac{1}{D^2 + D + 1}\sin 2x \qquad D^2 = -2^2 = -4$$

$$= \frac{1}{-4 + D + 1}\sin 2x$$

$$= \frac{1}{D-3}\sin 2x$$

Multiplying and dividing by $(D+3)$

$$= \frac{(D+3)\sin 2x}{(D+3)(D-3)}$$

$$= \frac{D(\sin 2x) + 3(\sin 2x)}{D^2 - 3^2}$$

$$= \frac{2\cos 2x + 3\sin 2x}{D^2 - 9} \qquad D^2 = -2^2 = -4$$

$$= \frac{2\cos 2x + 3\sin 2x}{-4-9}$$

$$= \frac{2\cos 2x + 3\sin 2x}{-13}$$

$$y = C.F. + P.I.$$

$$y = e^{-\frac{x}{2}}\left[C_1 \cos\frac{\sqrt{3}}{2}x + C_2 \sin\frac{\sqrt{3}}{2}x\right] - \frac{1}{13}(2\cos 2x + 3\sin 2x)$$

3. Solve: $(D^2 + 5D + 6)y = \cos x + e^{-2x}$

Solution Given equation is $(D^2 + 5D + 6)y = \cos x + e^{-2x}$

$$\text{A.E. is } m^2 + 5m + 6 = 0$$

$$(m+2)(m+3) = 0$$

$$m = -2, -3$$

$$C.F. = C_1e^{-2x} + C_2e^{-3x}$$

$$P.I. = \frac{1}{D^2+5D+6}\left(\cos x + e^{-2x}\right)$$

$$= \frac{\cos x}{D^2+5D+6} + \frac{e^{-2x}}{D^2+5D+6} = \text{P.I.}_1 + \text{P.I.}_2$$

$$\text{P.I.}_1 = \frac{\cos x}{D^2+5D+6} \qquad D^2 \to -1^2 = -1$$

$$= \frac{\cos x}{(-1)^2+5D+6} = \frac{\cos x}{5D+5}$$

$$= \frac{1}{5}\frac{\cos x}{D+1}$$

Multiplying and dividing by $(D-1)$

$$= \frac{1}{5}\frac{(D-1)\cos x}{(D+1)(D-1)}$$

$$= \frac{1}{5}\frac{-\sin x - \cos x}{(D^2-1^2)} \qquad D^2 \to -1^2 = -1$$

$$= \frac{1}{5}\frac{-\sin x - \cos x}{(-1-1)}$$

$$= \frac{1}{10}(\sin x + \cos x)$$

$$P.I._2 = \frac{e^{-2x}}{D^2+5D+6} \qquad D \to -2$$

$$= \frac{e^{-2x}}{(-2)^2+5(-2)+6} \qquad Dr = 0$$

Differentiate and multiply 'x'

$$= \frac{xe^{-2x}}{2D+5} \qquad D \to -2$$

$$= \frac{xe^{-2x}}{2(-2)+5}$$

$$= \frac{xe^{-2x}}{1} = xe^{-2x}$$

$$P.I. = P.I._1 + P.I._2$$

$$P.I. = \frac{1}{10}(\sin x + \cos x) + xe^{-2x}$$

$$y = C.F. + P.I.$$

$$y = C_1e^{-2x} + C_2e^{-3x} + \frac{1}{10}(\sin x + \cos x) + xe^{-2x}$$

4. Solve: $\left(D^2 + 3D + 2\right)y = \cos^2 x$

Solution Given equation is $\left(D^2 + 3D + 2\right)y = \cos^2 x$

$$\text{A.E. is } m^2 + 3m + 2 = 0$$

$$(m+1)(m+2) = 0$$

$$m = -1, -2$$

$$C.F. = C_1e^{-x} + C_2e^{-2x}$$

$$P.I. = \frac{1}{D^2 + 3D + 2}\cos^2 x$$

$$\text{where } \cos^2 x = \frac{1}{2}(1 + \cos 2x)$$

$$= \frac{1}{2}\left[\frac{1}{D^2 + 3D + 2}e^{0x} + \frac{1}{D^2 + 3D + 2}\cos 2x\right]$$

$$= \frac{1}{2}\left[P.I._1 + P.I._2\right]$$

$$P.I._1 = \frac{1}{D^2 + 3D + 2}e^{0x} \quad D \to 0$$

$$= \frac{1}{0 + 0 + 2}e^{0x} = \frac{1}{2}$$

$$P.I._2 = \frac{1}{D^2 + 3D + 2}\cos 2x \quad D^2 \to -2^2 = -4$$

$$= \frac{1}{-4 + 3D + 2}\cos 2x$$

$$= \frac{1}{3D - 2}\cos 2x$$

$$= \frac{(3D + 2)\cos 2x}{(3D - 2)(3D + 2)}$$

$$= \frac{-3\sin 2x(2) + 2\cos 2x}{(3D)^2 - 2^2}$$

$$= \frac{-6\sin 2x + 2\cos 2x}{9(-4)-4} \qquad D^2 \to -2^2 = -4$$

$$= \frac{-2(3\sin 2x - \cos 2x)}{-40}$$

$$= \frac{3\sin 2x - \cos 2x}{20}$$

$$\text{P.I.} = \frac{1}{2}\left[\frac{1}{2} + \frac{1}{20}(3\sin 2x - \cos 2x)\right]$$

$$y = C.F. + P.I.$$

$$y = C_1 e^{-x} + C_2 e^{-2x} + \frac{1}{2}\left[\frac{1}{2} + \frac{1}{20}(3\sin 2x - \cos 2x)\right]$$

or
$$y = C_1 e^{-x} + C_2 e^{-2x} + \frac{1}{4} + \frac{1}{40}(3\sin 2x - \cos 2x)$$

5. Solve: $(D^2 + 4)y = \sin^2 x$

Solution Given equation is $(D^2 + 4)y = \sin^2 x$

A.E. is $m^2 + 4 = 0$

$$m^2 = -4$$

$$m = \pm 2i$$

$$C.F. = C_1 \cos 2x + C_2 \sin 2x$$

$$P.I. = \frac{1}{D^2+4}\sin^2 x$$

where
$$\sin^2 x = \frac{1}{2}(1 - \cos 2x)$$

$$= \frac{1}{2}\left[\frac{1}{D^2+4}e^{0x} - \frac{1}{D^2+4}\cos 2x\right] = \frac{1}{2}[P.I._1 - P.I._2]$$

$$P.I._1 = \frac{1}{D^2+4}e^{0x} \qquad D \to 0$$

$$= \frac{1}{0+4}e^{0x} = \frac{1}{4}$$

$$P.I._2 = \frac{1}{D^2+4}\cos 2x \qquad D^2 \to -2^2 = -4$$

$$= \frac{1}{-4+4}\cos 2x \qquad Dr = 0$$

$$= \frac{1}{2D} x\cos 2x \times \frac{D}{D}$$

$$= \frac{-2x\sin 2x}{2D^2} \qquad D^2 \to -2^2 = -4$$

$$= \frac{-2x\sin 2x}{2(-4)} = \frac{x\sin 2x}{4}$$

$$P.I. = \frac{1}{2}\left[\frac{1}{4} - \frac{1}{4}x\sin 2x\right]$$

$$y = C.F. + P.I.$$

$$y = C_1\cos 2x + C_2\sin 2x + \frac{1}{2}\left[\frac{1}{4} - \frac{1}{4}x\sin 2x\right]$$

or $$y = C_1\cos 2x + C_2\sin 2x + \frac{1}{8} - \frac{x}{8}\sin 2x$$

6. Solve: $(D^3+1)y = \sin 3x - \cos^2\left(\frac{x}{2}\right)$

Solution Given equation is $(D^3+1)y - \sin 3x - \cos^2\left(\frac{x}{2}\right)$

$$\text{A.E. is } m^3 + 1 = 0$$

$$\text{i.e. } (m+1)(m^2 - m + 1) = 0$$

$$m + 1 = 0, \quad m^2 - m + 1 = 0$$

$$m = -1$$

$$m = \frac{-b \pm \sqrt{b^2 - 4ac}}{2a}$$

$$m = \frac{1 \pm \sqrt{1-4}}{2}$$

$$m = \frac{1 \pm \sqrt{3}i}{2}$$

$$m = -1\ ; \ m = \frac{1 \pm \sqrt{3}i}{2}$$

$$C.F. = C_1e^{-x} + \left(C_2\cos\frac{\sqrt{3}}{2}x + C_3\sin\frac{\sqrt{3}}{2}x\right)e^{\frac{1}{2}x}$$

$$P.I. = \frac{1}{D^3+1}\left[\sin 3x - \cos^2\left(\frac{x}{2}\right)\right]$$

$$\text{where } \cos^2\left(\frac{x}{2}\right) = \frac{1}{2}(1+\cos x)$$

$$= \frac{1}{D^3+1}\sin 3x - \frac{1}{2}\frac{1}{D^3+1}(1+\cos x)$$

$$= \frac{1}{D^3+1}\sin 3x - \frac{1}{2}\left[\frac{1}{D^3+1}e^{0x} + \frac{1}{D^3+1}\cos x\right]$$

$$= P.I._1 - \frac{1}{2}[P.I._2 + P.I._3]$$

$$P.I._1 = \frac{1}{D^3+1}\sin 3x$$

$$= \frac{1}{D^2\cdot D+1}\sin 3x \quad D^2 \to -3^2 = -9$$

$$= \frac{1}{-9D+1}\sin 3x$$

$$= \frac{1}{1-9D}\sin 3x \times \frac{1+9D}{1+9D}$$

$$= \frac{(1+9D)\sin 3x}{1-81D^2} \quad D^2 \to -3^2 = -9$$

$$= \frac{\sin 3x + 27\cos 3x}{1-81(-9)} = \frac{\sin 3x + 27\cos 3x}{730}$$

$$P.I._2 = \frac{1}{D^3+1}e^{0x} \quad D \to 0 \quad = \frac{1}{0+1}e^{0x} \quad = 1$$

$$P.I._3 = \frac{1}{D^3+1}\cos x$$

$$= \frac{1}{D^2 D+1}\cos x \quad D^2 \to -1^2 = -1$$

$$= \frac{\cos x}{-D+1} \times \frac{1+D}{1+D} \quad D^2 \to -1^2$$

$$= \frac{\cos x - \sin x}{D^2-1} = \frac{\cos x - \sin x}{-2}$$

$$P.I. = P.I._1 - \frac{1}{2}\left[P.I._2 + P.I._3\right]$$

$$= \frac{\sin 3x + 27\cos 3x}{730} - \frac{1}{2} - \frac{1}{4}(\cos x - \sin x)$$

$$y = C.F + P.I.$$

$$y = C_1 e^x + \left(C_2 \cos\frac{\sqrt{3}}{2}x + C_3 \sin\frac{\sqrt{3}}{2}x\right)e^{\frac{1}{2}x}$$

$$+ \frac{\sin 3x + 27\cos 3x}{730} - \frac{1}{2} - \frac{1}{4}(\cos x - \sin x)$$

7. Solve: $(D^2 + 9)y = \cos 2x \cdot \cos x$

Solution Given equation is $(D^2 + 9)y = \cos 2x \cdot \cos x$

A.E. is $m^2 + 9 = 0$

$$m^2 = -9$$

$$m = \pm 3i$$

$$C.F. = C_1 \cos 3x + C_2 \sin 3x$$

$$\text{P.I.} = \frac{1}{D^2 + 9}\cos 2x \cdot \cos x$$

where $\cos 2x \cdot \cos x = \frac{1}{2}(\cos x + \cos 3x)$

$$P.I. = \frac{1}{2}\frac{1}{D^2 + 9}(\cos x + \cos 3x)$$

$$= \frac{1}{2}\left[\frac{\cos x}{D^2 + 9} + \frac{\cos 3x}{D^2 + 9}\right] = \frac{1}{2}\left[P.I_2 + P.I._2\right]$$

$$P.I._1 = \frac{\cos x}{D^2 + 9} \quad D^2 \to -1^2 = -1$$

$$= \frac{1}{-1 + 9}e^{0x} = \frac{1}{8}\cos x$$

$$P.I._2 = \frac{1}{D^2 + 4}\cos 3x \quad D^2 \to -3^2 = -9$$

$$= \frac{1}{-9 + 9}\cos 2x \quad Dr = 0$$

$$= \frac{1}{2D}x\cos 3x \times \frac{D}{D}$$

$$= \frac{-3x\sin 3x}{2D^2} \qquad D^2 \to -3^2 = -9$$

$$= \frac{-3x\sin 3x}{2(-9)} = \frac{x\sin 3x}{6}$$

$$P.I. = \frac{1}{2}\left[\frac{1}{8}\cos x + \frac{1}{6}x\sin 3x\right]$$

$$y = C.F. + P.I.$$

$$y = C_1\cos 2x + C_2\sin 2x + \frac{1}{2}\left[\frac{1}{8}\cos x + \frac{1}{6}x\sin 3x\right]$$

$$\text{or } y = C_1\cos 3x + C_2\sin 3x + \frac{1}{16}\cos x + \frac{x}{12}\sin 3x$$

EXERCISES

Solve the following equations

1. $(D^2+1)y = \sin 2x$ $\left[\textbf{Ans}: y = C_1\cos x + C_2\sin x - \frac{1}{3}\sin 2x\right]$
2. $(D_2+4)y = \sin 2x + \cos 3x$

$\left[\textbf{Ans}: y = C_1e^{2x} + C_2{}^{-2x} - \frac{1}{8}\sin 2x - \frac{1}{13}\cos 3x\right]$

3. $(D^2+9)y = \sin 3x$ $\left[\textbf{Ans}: y = C_1\cos 3x + C_2\sin 3x - \frac{x}{6}\cos 3x\right]$
4. $(D^2+16)y = \cos 4x$ $\left[\textbf{Ans}: y = C_1\cos 4x + C_2\sin 4x + \frac{x}{8}\sin 4x\right]$
5. $(D^2+1)y = \sin x\sin 2x$

$\left[\textbf{Ans}: y = C_1\cos x + C_2\sin x + \frac{1}{16}(4x\sin + \cos 3x)\right]$

6. $(D^2-2D-8)y = 4\cos 2x$

$\left[\textbf{Ans}: y = C_1e^{4x} + C_2e^{-2x} - \frac{1}{10}(3\cos 2x + \sin 2x)\right]$

8. $(D^2+3D+2)y = 4\cos^2 x$

$\left[\textbf{Ans}: y = C_1e^{-x} + C_2e^{-2x} + 1 + \frac{1}{10}(3\sin 2x - \cos 2x)\right]$

9. $(D^2-4)y = \cos x\cos 2x$

$\left[\textbf{Ans}: \left(C_1e^{2x} + C_2e^{-2x} - \frac{1}{2}\left(\frac{1}{13}\cos 3x + \frac{1}{5}\cos x\right)\right)\right]$

10. $\left(D^2-4D+4\right)y=\sin 2x+\cos 2x$

$$\left[\textbf{Ans: } y=(C_1+C_2x)e^{2x}+\frac{1}{8}(\cos 2x-\sin 2x)\right]$$

11. $\left(D^2+8D+25\right)y=48\cos x-16\sin x$

$$\left[\textbf{Ans: } y=e^{-4x}\left(C_1\cos 3x+C_2\sin 3x\right)+2\cos x\right]$$

12. $\left(D^3+D^2+D+1\right)y=\sin 3x$

$$\left[\textbf{Ans: } y=C_1e^{-x}+C_2\cos x+C_3\sin x+\frac{1}{80}(3\cos 3x-\sin 3x)\right]$$

13. $\left(D^4-2D^2+1\right)y=\cos x$

$$\left[\textbf{Ans: } y=\left(C_1+C_2x\right)e^{x}+\left(C_3+C_4x\right)+\frac{1}{4}\cos x\right]$$

14. $\left(D^3+8\right)y=\sin 2x$

$$\left[\textbf{Ans: } y=C_1e^{-2x}+e^{x}\left(C_2\cos\sqrt{3}+C_3\sin\sqrt{3x}\right)+\frac{1}{16}(\cos 2x+\sin 2x)\right]$$

15. $\left(D^4-16\right)y=\sin x\cos x$

$$\left[\textbf{Ans: } y=C_1e^{2x}+C_2e^{-2x}+C_3\cos 2x+C_4\sin 2x+\frac{1}{64}x\cos 2x\right]$$

16. $(D-1)^3\,y=\cos^2 x$

$$\left[\textbf{Ans: } y=\left(C_1+C_2x+C_3x^2\right)e^{x}-\frac{1}{2}-\frac{1}{250}(2\sin 2x-11\cos 2x)\right]$$

17. $\left(D^3-3D+2\right)y=\sin 2x$

$$\left[\textbf{Ans: } \left(C_1+C_2x\right)e^{x}+C_3e^{-2x}+\frac{1}{100}(7\cos 2x+\sin 2x)\right]$$

18. $\left(D^2+1\right)\left(D^2+4\right)y=\cos 2x+\sin x$

$$\left[\textbf{Ans: } y=C_1\cos x+C_2\sin x+C_3\sin x+C_3\cos 2x+C_4\sin 2x-\frac{1}{12}x\sin 2x-\frac{1}{6}x\cos x\right]$$

19. $\dfrac{d^3y}{dx^3}+y=65\cos(2x+1)+e^{-x}$

$$\left[\textbf{Ans: } C_1e^{-x}+e^{\frac{x}{2}}\left\{C_2\cos\frac{\sqrt{3}}{2}x+C_3\sin\frac{\sqrt{3}}{2}x\right\}+\cos(2x+1)-8\sin(2x+1)+\frac{1}{3}xe^{-x}\right]$$

Type 3

1. Solve: $y'' + 3y' + 2y = 12x^2$

Solution Given equation is $(D^2 + 3D + 2)y = 12x^2$

$$\text{A.E. is } m^2 + 3m + 2 = 0$$

$$\text{i.e., } (m+1)(m+2) = 0$$

$$m = -1, -2$$

$$\text{C.F.} = C_1 e^{-x} + C_2 e^{-2x}$$

$$P.I. = \frac{12x^2}{D^2 + 3D + 2}$$

we need to divide for obtaining the P.I.

Note: $3D\,(6x^2) = 36x$

$D^2\,(6x^2) = 12$

$$\begin{array}{r|l} & 6x^2 - 18x + 21 \\ \hline 2 + 3D + D^2 & 12x^2 \\ & 12x^2 + 36x + 12 \\ \hline & -36x - 12 \\ & -36x - 54 \\ \hline & 42 \\ & 42 \\ \hline & 0 \end{array}$$

Note : $3D(6x^2) = 36x$

$D^2(6x^2) = 12$

Hence $\quad$ P.I $= 6x^2 - 18x + 21$

$\therefore$ The general solution is

$$y = \text{C.F} + \text{P.I}$$

$$y = C_1 e^{-x} + C_2 e^{-2x} + (6x^2 - 18x + 21)$$

2. Solve: $\dfrac{d^2y}{dx^2} + 2\dfrac{dy}{dx} + y = 2x + x^2$

Solution Given equation is $(D^2 + 2D + 1)y = 2x + x^2$

$$\text{A.E. is } m^2 + 2m + 1 = 0$$

i.e. $\quad (m + 1)^2 = 0$

i.e. $\quad (m + 1)(m + 1) = 0$

$$m = -1, -1$$

$$\text{C.F} = (C_1 + C_2x)e^{x}$$

$$\text{P.I.} = \frac{2x + x^2}{D^2 + 2D + 1}$$

$$\text{P.I} = \frac{x^2 + 2x}{1 + 2D + D^2}$$

we need to divide for obtaining the P.I.

$$\begin{array}{r|l}
 & x^2 - 2x + 2 \\
\hline
1 + 2D + D^2 & x^2 + 2x \\
 & x^2 + 4x + 2 \\
\hline
 & -2x - 2 \\
 & -2x - 4 \\
\hline
 & 2 \\
 & 2 \\
\hline
 & 0
\end{array}$$

Hence $\qquad \text{P.I} = x^2 - 2x + 2$

$\therefore$ The general solution is

$$y = \text{C.F} + \text{P.I}$$

$$y = (C_1 + C_2x)e^{x} + (x^2 - 2x + 2)$$

3. Solve: $\dfrac{d^2y}{dx^2} + 5\dfrac{dy}{dx} + 6y = x^2$

Solution Given equation is $(D^2 + 5D + 6)y = x^2$

$$\text{A.E. is } m^2 + 5m + 6 = 0$$

i.e. $\qquad (m + 2)(m + 3) = 0$

$$m = -2, -3$$

$$\text{C.F} = C_1e^{-2x} + C_2e^{-3x}$$

$$\text{P.I} = \frac{x^2}{D^2 + 5D + 6} = \frac{x^2}{6 + 5D + D^2}$$

we need to divide for obtaining the P.I.

$$
\begin{array}{r|l}
 & x^2 - \frac{5x}{18} + \frac{19}{108} \\
\hline
6 + 5D + D^2 & x^2 \\
 & x^2 + \frac{5x}{18} + \frac{1}{3} \\
\hline
 & -\frac{5x}{3} - \frac{1}{3} \\
 & -\frac{5x}{3} - \frac{25}{18} \\
\hline
 & 19/18 \\
 & 19/18 \\
\hline
 & 0
\end{array}
$$

Note : $5D\left(\frac{x^2}{6}\right) = \frac{5x}{3}$

$D^2\left(\frac{x^2}{6}\right) = \frac{1}{3}$

$5D\left(\frac{-5x}{18}\right) = \frac{-25}{18}$

Hence $\quad \text{P.I} = x^2 - \frac{5x}{18} + \frac{19}{108}$

$\therefore$ The general solution is

$$y = \text{C.F} + \text{P.I}$$

$$y = C_1 e^{-2x} + C_2 e^{-3x} + \left(x^2 - \frac{5x}{18} + \frac{19}{108}\right)$$

or $$y = C_1 e^{-2x} + C_2 e^{-3x} + \frac{1}{108}(18x^2 - 30x + 19)$$

4. Solve: $(D^3 + 2D^2 + D)y = x^3$

Solution Given equation is $(D^3 + 2D^2 + D)y = x^3$

A.E. is $m^3 + 2m^2 + m = 0$

$\Rightarrow \quad m(m^2 + 2m + 1) = 0$

i.e. $\quad m(m + 1)^2 = 0$

i.e. $\quad m(m + 1)(m + 1) = 0$

$$m = 0, -1, -1$$

$$\text{C.F} = C_1 + (C_2 + C_3 x)e^{-x}$$

$$\text{P.I} = \frac{x^3}{D + 2D^2 + D^3}$$

we need to divide for obtaining the P.I.

$$
\begin{array}{r|l}
 & \frac{x^4}{4} - 2x^3 + 9x^2 - 24x \\ \hline
D + 2D^2 + D^3 & x^3 \\
 & x^3 + 6x^2 + 6x \\ \hline
 & -6x^2 - 6x \\
 & -6x^2 - 24x - 12 \\ \hline
 & 18x + 12 \\
 & 18x + 36 \\ \hline
 & -24 \\
 & -24 \\ \hline
 & 0
\end{array}
$$

Note : $\frac{x^3}{D} = \int x^3 dx = \frac{x^4}{4}$

$$2D^2\left(\frac{x^4}{4}\right) = 6x^2$$

$$D^3\left(\frac{x^4}{4}\right) = 6x$$

$$\frac{-6x^2}{D} = \int -6x^2 - dx = -2x^3$$

$$\frac{18x}{D} = \int 18x dx = 9x^2$$

$$\frac{-12}{D} = \int (-24) dx = -24x$$

Hence $\quad \text{P.I} = \frac{x^4}{4} - 2x^3 + 9x^2 - 24x$

$\therefore$ The general solution is

$$y = \text{C.F} + \text{P.I}$$

$$y = C_1 + (C_2 + C_3 x)e^{-x} + \left(\frac{x^4}{4} - 2x^3 + 9x^2 - 24x\right)$$

EXERCISES

1. $(D^2 - 9)y = 2x - 1$ $\quad\left[\textbf{Ans}: C_1e^{3x} + C_2e^{-3x} - \frac{1}{9}(2x-1)\right]$

2. $(D^2 + 4)y = x^2 - x$ $\quad\left[\textbf{Ans}: C_1\cos 2x + C_2 \sin 2x + \frac{1}{8}(2x^2 - 2x - 1\right]$

3. $(D^2 - 4D + 4)y = x^2$ $\quad\left[\textbf{Ans}: y = (C_1 + C_2x)e^{2x} + \frac{1}{8}(2x^2 + 4x + 3\right]$

4. $(D^2 + D - 6)y = x$ $\quad\left[\textbf{Ans}: y = C_1e^{2x} + C_2e^{-3x} - \frac{1}{36}(6x+1)\right]$

5. $(D^2 + 2D + 1)y = x^2 + 2x$ $\quad\left[\textbf{Ans}: y = (C_1 + C_2x)e^{-x} + x^3 - 2x + 2\right]$

6. $(D^2-2D+3)y=x^2$

$$\left[\textbf{Ans: } e^x\left(C_1\cos\sqrt{2}x+C_2\sin\sqrt{2}x\right)+\frac{1}{27}\left(9x^2+12x+2\right)\right]$$

7. $\left(D^2-5D+6\right)y=x^2+x-2$

$$\left[\textbf{Ans: } C_1e^{2x}+C_2e^{3x}+\frac{1}{54}\left(9x^2+12x+2\right)\right]$$

8. $(D^3+1)y=x^3$

$$\left[\textbf{Ans: } y=C_1e^{-x}+e^{x/2}\left(C_2\cos\frac{\sqrt{3}}{2}+C_3\sin\frac{\sqrt{3}}{2}x\right)+x^3-6\right]$$

9. $\left(D^3-D^2\right)y=x^2-3x+1$ $\quad\left[\textbf{Ans: } y=C_1+C_2x+C_3e^x-\frac{x^3}{12}(x-2)\right]$

10. $\left(D^3+8\right)y=x^4+2x+1$

$$\left[\textbf{Ans: } y=C_1e^{-2x}+e^x\left(C_2\cos\sqrt{3}x+C_3\sin\sqrt{3}x\right)+\frac{1}{8}\left(x^4-x+1\right)\right]$$

Type 4

1. Solve: $\dfrac{d^2y}{dx^2}+2\dfrac{dy}{dx}-3y=e^x\cos x$

Solution We have

$$(D^2+2D-3)\,y = e^x\cos x$$

$$\text{A.E. is } m^2+2m-3 = 0$$

$$\text{i.e., } (m+3)(m-1) = 0$$

$$m = -3,\ 1$$

$$\text{C.F.} = C_1e^{-3x}+C_2e^x$$

$$P.I. = \frac{1}{D^2+2D-3}e^x\cos x$$

Taking e^x outside the operator and changing D to $D+1$

$$= e^x\frac{1}{(D+1)^2+2(D+1)-3}\cos x$$

$$= e^x\frac{1}{D^2+4D}\cos x \quad (D^2\to -1^2)$$

$$= e^x\frac{1}{-1+4D}\cos x$$

$$= e^x\left[\frac{\cos x}{4D-1}\times\frac{4D+1}{4D+1}\right]$$

$$= e^x\left[\frac{-4\sin x+\cos x}{16D^2-1}\right] \quad (D^2 \to -1^2)$$

$$= e^x\left[\frac{-4\sin x+\cos x}{-17}\right]$$

$$= \frac{e^x}{17}(4\sin x-\cos x)$$

$$y = \text{C.F} + \text{P.I}$$

$$y = C_1e^{-3x} + C_2e^x + \frac{e^x}{17}(4\sin x-\cos x)$$

2. Solve: $(D^3 + 1)\, y = 5e^x x^2$

Solution A.E. is

$$m^3 + 1 = 0$$

$$\text{i.e., } (m+1)(m^2-m+1) = 0$$

$$(m+1) = 0,\quad m^2 - m + 1 = 0,$$

$$m = -1,\ m = \frac{1\pm\sqrt{3}i}{2}$$

$$\text{C.F.} = C_1e^{-x} + e^{x/2}\left(C_2\cos\frac{\sqrt{3}}{2}x + C_3\sin\frac{\sqrt{3}}{2}x\right)$$

$$\text{P.I} = \frac{1}{D^3+1}5e^x x^2$$

Taking e^x outside the operator and changing D to $D + 1$

$$= e^x\frac{1}{(D+1)^3+1}5x^2$$

$$= e^x\frac{5x^2}{D^3+3D^2+3D+2}$$

$$= \frac{5e^x}{2}\left[\frac{2x^2}{2+3D+3D^2+D^3}\right]$$

(For a convenient division we have multiplied and divided by 2)

$$
\begin{array}{r|l}
 & x^2-3x+\frac{3}{2} \\
\hline
2+3D+3D^2+D^3 & 2x^2 \\
 & 2x^2+6x+6 \\
\hline
 & -6x-6 \\
 & -6x-9 \\
\hline
 & 3 \\
 & 3 \\
\hline
 & 0
\end{array}
$$

$$\text{P.I.} = \left(x^2-3x+\frac{3}{2}\right)\cdot\frac{5e^x}{2}$$

$$= \frac{5e^x}{4}\left(2x^2-6x+3\right)$$

$$y = \text{C.F} + \text{P.I}$$

$$y = C_1e^{-x}+e^{x/2}\left(C_2\cos\frac{\sqrt{3}}{2}x+C_3\sin\frac{\sqrt{3}}{2}x\right)$$

$$+\frac{5e^x}{4}\left(2x^2-6x+3\right)$$

3. Solve: $(D^3 - 4D + 3)\,y = e^{2x}\sin 3x$

Solution A.E. is

$$m^2 - 4m + 3 = 0$$

$$\text{i.e., } (m-1)(m-3) = 0$$

$$m = 1, 3$$

$$\text{C.F.} = C_1e^x + C_2e^{3x}$$

$$\text{P.I} = \frac{1}{D^3-4D+3}e^{2x}\sin 3x$$

Taking e^{2x} outside the operator and changing D to $D + 2$

$$= e^{2x}\frac{1}{(D+2)^2-4(D+2)+3}\sin 3x$$

$$= e^{2x}\left[\frac{1}{D^2-1}\right]\sin 3x \quad (D^2 \to -3^2)$$

$$= -\frac{1}{10}e^{2x}\sin 3x$$

$$y = \text{C.F} + \text{P.I}$$

$$y = C_1e^x + C_2e^{3x} - \frac{1}{10}e^{2x}\sin 3x$$

4. Solve: $\dfrac{d^2y}{dx^2} + 4y = 2e^x \sin^2 x$

Solution We have

$$(D^2 + 4)y = 2e^x \sin^2 x$$

$$\text{A.E., } (m^2 + 4) = 0$$

$$m^2 = -4$$

$$m = \pm 2i$$

$$\text{C.F.} = C_1 \cos 2x + C_2 \sin 2x$$

$$\text{P.I} = \frac{1}{D^2+4} 2e^x \sin^2 x$$

where $\sin^2 x = \dfrac{1-\cos 2x}{2}$

$$= \frac{1}{D^2+4} e^x (1 - \cos 2x)$$

$$= \frac{1}{D^2+4} (e^x - e^x \cos 2x)$$

$$= \frac{e^x}{D^2+4} - \frac{e^x \cos 2x}{D^2+4}$$

$$= \text{P.I.}_1 - \text{P.I.}_2$$

$$\text{P.I.}_1 = \frac{1}{D^2+4} e^x \quad (D \to 1)$$

$$= \frac{e^x}{5}$$

$$\text{P.I.}_2 = \frac{e^x \cos 2x}{D^2+4}$$

Taking e^x outside the operator and changing D to $D + 1$

$$= e^x \frac{1}{(D+1)^2+4} \cos 2x$$

$$= e^x \left[\frac{1}{D^2+2D+1+4}\right] \cos 2x$$

$$= e^x \left[\frac{1}{D^2+2D+5}\right] \cos 2x \qquad (D^2 \to -2^2)$$

$$= e^x \frac{1}{-2^2+2D+5} \cos 2x = e^x \frac{1}{2D+1} \cos 2x \times \frac{2D-1}{2D-1}$$

$$= e^x \left[\frac{-4\sin 2x - \cos 2x}{4D^2-1}\right] \qquad (D^2 \to -2^2)$$

$$= e^x \left[\frac{-4\sin 2x - \cos 2x}{-17}\right]$$

$$= \frac{e^x}{17}(4\sin 2x + \cos 2x)$$

$$\therefore \quad \text{P.I.} = \frac{e^x}{5} + \frac{e^x}{17}(4\sin 2x + \cos 2x)$$

$$y = \text{C.F} + \text{P.I}$$

$$y = C_1 \cos 2x + C_2 \sin 2x + \frac{e^x}{5} + \frac{e^x}{17}(4\sin 2x + \cos 2x)$$

5. Solve: $(D^2 - 4D + 1)\, y = 2x\, e^{3x}$

Solution A.E. is

$$m^2 - 4m + 3 = 0$$

$$\text{i.e., } (m-1)(m-3) = 0$$

$$m = 1, 3$$

$$\text{C.F.} = C_1 e^x + C_2 e^{3x}$$

$$\text{P.I} = \frac{1}{D^2 - 4D + 3} 2xe^{3x}$$

Taking e^{3x} outside the operator and changing D to $D + 3$

$$= e^{3x} \frac{2x}{(D+3)^2 - 4(D+3) + 3}$$

$$= e^{3x} \left[\frac{2x}{D^2+2D}\right]$$

By division method

$$\begin{array}{r|l} & x^2/2 - x/2 \\ \hline 2D + D^2 & 2x \\ & 2x + 1 \\ \hline & -1 \\ & -1 \\ \hline & 0 \end{array}$$

Note : $\frac{2x}{2D} = \int x dx = \frac{x^2}{2}$,

$$\frac{-1}{2D} = \int \frac{-1}{2} dx = -\frac{x}{2}$$

$$\text{P.I.} = e^{3x}\left(\frac{x^2}{2} - \frac{x}{2}\right)$$

$$y = \text{C.F} + \text{P.I}$$

$$y = C_1 e^x + C_2 e^{3x} + e^{3x}\left(\frac{x^2}{2} - \frac{x}{2}\right)$$

Type 5

1. Solve: $\frac{d^2 y}{dx^2} + 4y = x \sin x$

Solution We have

$$(D^2 + 4)\, y = x \sin x$$

$$\text{A.E. is } m^2 + 4 = 0$$

$$m^2 = -4$$

$$m = \pm 2i$$

$$\text{C.F.} = C_1 \cos 2x + C_2 \sin 2x$$

$$P.I. = \frac{1}{D^2 + 4} x \sin x$$

Let us use $\frac{xV}{f(D)} = \left[x - \frac{f'(D)}{f(D)}\right]\frac{V}{f(D)}$

$$\frac{x \sin x}{D^2 + 4} = \left[x - \frac{2D}{D^2 + 4}\right]\frac{\sin x}{D^2 + 4} \qquad (D^2 \to -1^2)$$

$$= \frac{x \sin x}{D^2 + 4} - \frac{2D(\sin x)}{(D^2 + 4)^2} \qquad (D^2 \to -1^2)$$

$$= \frac{x \sin x}{3} - \frac{2 \cos x}{3^2}$$

$$= \frac{x \sin x}{3} - \frac{2 \cos x}{9}$$

$$\text{P.I} = \frac{1}{9}(3x\sin x - 2\cos x)$$

$$y = \text{C.F} + \text{P.I.}$$

$$= C_1\cos 2x + C_2\sin 2x + \frac{1}{9}(3x\sin x - 2\cos x)$$

2. Solve: $(D^2 + 2D + 1)y = x\cos x$

Solution A.E. is

$$m^2 + 2m + 1 = 0$$

i.e., $$(m+1)^2 = 0$$

$$m = -1, -1$$

$$\text{C.F.} = (C_1 + C_2 x)e^{-x}$$

$$P.I. = \frac{x\cos x}{D^2 + 2D + 1}$$

Let us we have $$\frac{xV}{f(D)} = \left[x - \frac{f'(D)}{f(D)}\right]\frac{V}{f(D)}$$

$$= \left[x - \frac{2D+2}{D^2+2D+1}\right]\frac{\cos x}{D^2+2D+1}$$

$$= \frac{x\cos x}{D^2+2D+1} - \frac{(2D+2)\cos x}{(D^2+2D+1)^2}$$

$$= \text{P.I}_1 - \text{P.I}_2$$

$$\text{P.I.}_1 = \frac{x\cos x}{D^2+2D+1} \qquad (D^2 \to -1^2)$$

$$= \frac{x\cos x}{-1+2D+1} = \frac{x\cos x}{2D}$$

$$= \frac{x\cos x \times D}{2D \times D}$$

$$= \frac{-x\sin x}{2D^2} \qquad (D^2 \to -1^2)$$

$$\text{P.I.}_1 = \frac{x}{2}\sin x$$

$$\text{P.I.}_2 = \frac{(2D+2)\cos x}{(D^2+2D+1)^2} \qquad (D^2 \to -1^2)$$

$$= \frac{-2\sin x + 2\cos x}{(2D)^2}$$

$$= \frac{-2\sin x + 2\cos x}{4D^2}$$

$$= \frac{2\sin x - 2\cos x}{4}$$

$$= \frac{1}{2}(\sin x - \cos x)$$

$$\text{P.I.} = \frac{1}{2}x\sin x - \frac{1}{2}(\sin x - \cos x)$$

$$= \frac{1}{2}(x\sin x - \sin x + \cos x)$$

$$y = \text{C.F} + \text{P.I.}$$

$$= (C_1 + C_2x)e^{-x} + \frac{1}{2}(x\sin x - \sin x + \cos x)$$

3. Solve: $\frac{d^2y}{dx^2} - y = xe^x \sin x$

Solution We have the equation $(D^2 - 1)y = x e^x \sin x$

A.E. is $m^2 - 1 = 0$

i.e., $m^2 = 1$

$$m = +1, -1$$

$$\text{C.F.} = C_1e^x + C_2e^{-x}$$

$$P.I. = \frac{1}{D^2 - 1}xe^x \sin x$$

Taking e^x outside the operator and changing D to $D + 1$

$$= e^x \frac{1}{(D+1)^2 - 1}x\sin x$$

$$= e^x \left[\frac{x\sin x}{D^2 + 2D}\right]$$

Let us we have $\frac{xV}{f(D)} = \left[x - \frac{f'(D)}{f(D)}\right]\frac{V}{f(D)}$

$$= e^x \left[x - \frac{2D+2}{D^2 + 2D}\right] \cdot \frac{\sin x}{D^2 + 2D} \qquad (D^2 \to -1^2)$$

$$= e^x\left[\frac{x\sin x}{D^2+2D}-\frac{2(D+1)}{D^2(D+2)^2}\sin x\right]$$

$$= e^x\left[\frac{x\sin x}{2D-1}-\frac{2(D+1)}{D^2(D+2)^2}\sin x\right]$$

$$= e^x\left[\frac{x(2D+1)\sin x}{4D^2-1}-\frac{2(D+1)}{-1(D^2+4D+4)^2}\sin x\right]$$

$$= e^x\left[\frac{2x\cos x+x\sin x}{-5}+\frac{2(D+1)(4D-3)}{16D^2-9}\sin x\right]$$

$$= e^x\left[\frac{-1}{5}x(2\cos x+\sin x)-\frac{2}{25}(4D^2+D-3)\sin x\right]$$

$$= e^x\left[\frac{-1}{5}x(2\cos x+\sin x)-\frac{2}{25}(-4\sin x+\cos x-3\sin x)\right]$$

$$\text{P.I.} = \frac{-1}{25}e^x[(5x-14)\sin x+2(5x+1)\cos x]$$

$\therefore$ The general solution is

$$y = \text{C.F} + \text{P.I.}$$

$$= C_1e^x+C_2e^{-x}-\frac{1}{25}e^x[(5x-14)\sin x+2(5x+1)\cos x]$$

4. Solve: $\dfrac{d^2y}{dx^2}-4\dfrac{dy}{dx}+4y=3x^2e^{2x}\sin 2x$

Solution We have the

$$(D^2-4D+4)y = 3x^2e^{2x}\sin 2x$$

A.E. is $m^2-4m+4 = 0$

i.e., $(m-2)^2 = 0$

$$m = 2, 2$$

$$\text{C.F.} = (C_1+C_2x)e^{2x}$$

$$P.I. = \frac{1}{(D-2)^2}3x^2e^{2x}\sin 2x$$

Taking e^{2x} outside the operator and changing D to $D+2$

$$= 3e^{2x}\frac{1}{D^2}x^2\sin 2x$$

We shall find $\left(\frac{1}{D^2}x^2\sin 2x\right)$ integrating twce

$$\frac{1}{D^2}x^2\sin 2x = \int x^2\sin 2x\,dx$$

Applying integration by parts

$$= x^2\left(\frac{-\cos 2x}{2}\right) - 2x\left(\frac{-\sin 2x}{4}\right) + 2\left(\frac{\cos 2x}{8}\right)$$

$$= \frac{-x^2\cos 2x}{2} + \frac{x\sin 2x}{2} + \frac{\cos 2x}{4}$$

Hence

$$\frac{1}{D^2}(x^2\sin 2x) = \frac{-1}{2}\int x^2\cos 2x\,dx + \frac{1}{2}\int x\sin 2x\,dx - \frac{1}{4}\int\cos 2x\,dx$$

$$= \frac{-1}{2}\left[x^2\frac{\sin 2x}{2} - 2x\left(\frac{-\cos 2x}{4}\right) + 2\left(\frac{-\sin 2x}{8}\right)\right]$$

$$+\frac{1}{2}\left[x\left(\frac{-\cos 2x}{2}\right) - 1\left(\frac{-\sin 2x}{4}\right)\right] + \frac{1}{4}\frac{\sin 2x}{2}$$

$$= \frac{-1}{8}(2x^2\sin 2x + 4x\cos 2x - 3\sin 2x)$$

$$\text{P.I.} = \frac{-3}{8}e^{2x}(2x^2\sin 2x + 4x\cos 2x - 3\sin 2x)$$

$\therefore$ The general solution is

$$y = \text{C.F} + \text{P.I.}$$

$$= (C_1 + C_2x)e^{2x}$$

$$-\frac{3}{8}e^{2x}(2x^2\sin 2x + 4x\cos 2x - 3\sin 2x)$$

5. Solve: $\frac{d^2y}{dx^2} - 2\frac{dy}{dx} + y = xe^x\sin x$

Solution We have the

$$(D^2 - 2D + 1)y = xe^x\sin x$$

A.E. is $m^2 - 2m + 1 = 0$

i.e., $(m-1)^2 = 0$

$$m = 1, 1$$

$$\text{C.F.} = (C_1 + C_2x)e^x$$

$$P.I. = \frac{1}{D^2 - 2D + 1} xe^x \sin x$$

$$= \frac{1}{(D-1)^2} xe^x \sin x \qquad (D \to D+1)$$

$$= e^x \frac{1}{D^2} x \sin x$$

Now $\frac{1}{D^2}(x \sin x) = \frac{1}{D}\int x \sin x \, dx$

Since $\frac{1}{D}$ is the integration

$$= \frac{1}{D}[x(-\cos x) - 1(-\sin x)]$$

On integration by parts

$$= \frac{1}{D}(-x\cos x + \sin x)$$

$$= -\int x \cos x \, dx + \int \sin x \, dx$$

$$= -\{x \sin x - 1(-\cos x)\} - \cos x$$

$$= -x \sin x - 2 \cos x$$

$$= -(x \sin x + 2 \cos x)$$

Hence P.I. $= -e^x (x \sin x + 2 \cos x)$

∴ The general solution is

$$y = \text{C.F} + \text{P.I.}$$

$$= (C_1 + C_2 x)e^x - e^x(x \sin x + 2\cos x)$$

EXERCISES

Solve the following equation

1. $(D^2 + 9)y = x \cos x$

$$\left[\textbf{Ans: } y = C_1 \cos 3x + C_2 \sin 3x + \frac{1}{32}(4x \cos x + \sin x)\right]$$

2. $(D^2 - 2D + 1)\, y = x \sin x$

$$\left[\textbf{Ans: } y = (C_1 + C_2 x)e^{-x} + \frac{1}{2}(\sin x + \cos x - 1\right]$$

3. $(D^3 - 1)y = x \sin 3x$

$$\left[\textbf{Ans: } y = C_1 e^x + C_2 e^{-x} - \frac{1}{50}(5x \sin 3x + 3\cos 3x)\right]$$

4. $(D^2 - 3D + 2)y = x \cos 2x$

$$\left[\textbf{Ans: } y = C_1e^x + C_2e^{2x} - \frac{1}{20}x(3\sin 2x + \cos 2x) - \frac{1}{200}(7\sin 2x + 24\cos 2x)\right]$$

5. $\dfrac{d^2y}{dx^2} + a^2y = x\cos ax$

$$\left[\textbf{Ans: } y = C_1\cos ax + C_2\sin ax + \frac{1}{4a^2}(ax^2\sin ax + x\cos ax)\right]$$

6. $\dfrac{d^2y}{dx^2} - y = x^2\cos x$

$$\left\{\textbf{Ans: } y = C_1e^x + C_2e^{-x} - \frac{1}{2}\left[(x^2 - 1)\cos x - 2x\sin x\right]\right\}$$

7. $(D^2 - 4D + 4)y = 4x^2e^{2x}\cos 2x$

$$\left\{\textbf{Ans: } y = (C_1 + C_2x)e^{2x} - \frac{1}{2}e^{2x}\left[(2x^2 - 3)\cos 2x - 4x\sin 2x\right]\right\}$$

8. $\dfrac{d^4y}{dx^4} - y = x\sin x$

$$\left\{\textbf{Ans: } y = C_1\cos x + C_2\sin x + C_1e^x + C_4e^{-x} + \frac{1}{8}\left[(x^2 + 2)\cos 5x - 3x\sin x\right]\right\}$$